SpringerBriefs in Electrical and Computer Engineering

More information about this series at http://www.springer.com/series/10059

Kan Zheng • Lin Zhang • Wei Xiang • Wenbo Wang

Heterogeneous Vehicular Networks

Springer

Kan Zheng
School of Information and Communication
  Engineering
Beijing University of Posts
  and Telecommunications
Haidian District, Beijing, China

Wei Xiang
College of Science, Technology
  and Engineering
Division of Tropical Environments
  and Societies
James Cook University, Cairns
QLD, Australia

Lin Zhang
School of Information
  and Communication Engineering
Beijing University of Posts
  and Telecommunications
Haidian District, Beijing, China

Wenbo Wang
School of Information
  and Communication Engineering
Beijing University of Posts
  and Telecommunications
Haidian District, Beijing, China

ISSN 2191-8112   ISSN 2191-8120   (electronic)
SpringerBriefs in Electrical and Computer Engineering
ISBN 978-3-319-25620-7   ISBN 978-3-319-25622-1   (eBook)
DOI 10.1007/978-3-319-25622-1

Library of Congress Control Number: 2016930306

Printed on acid-free paper

This Springer imprint is published by Springer Nature
The registered company is Springer International Publishing AG Switzerland

Preface

With the advent of the intelligent transport system (ITS), vehicular communications networks have been widely studied in recent years. Dedicated short-range communications (DSRC) can provide efficient real-time information exchange between vehicles even with the lack of pervasive roadside communications infrastructure. Although mobile cellular networks are capable of providing great coverage for vehicular users, the requirement of stringent real-time safety services cannot always be guaranteed in mobile networks. Therefore, the Heterogeneous Vehicular NETwork (HetVNET), which integrates cellular networks with DSRC, emerges as a promising solution to meet the communications requirements of the ITS. Although there exist extensive reported studies on either DSRC or cellular networks, the combination of these two popular techniques remains a relatively nascent field of research. Building such HetVNETs requires thorough investigations into heterogeneity and its associated challenges.

The objective of this monograph is to present architectures of the HetVNET and to examine recent advances in Medium access control (MAC) layer designs for such systems. In Chap. 1, we present the motivation to the development of HetVNETs after a brief introduction to existing vehicular networks as well as the user cases and requirements of ITS services. Chapter 2 proposes an HetVNET architecture that utilizes a variety of wireless networking techniques, followed by the descriptions of various applications in some typical scenarios. Chapter 3 focuses on the MAC mechanisms of vehicular communications including a novel location-based channel congestion control mechanism. In order to well exploit the radio resources in HetVNETs, efficient resource allocation schemes are desired. Thus, not only the content-based scheme but also the cooperative one are presented in Chap. 4, following a short brief to the state-of-the-art. Finally, Chap. 5 suggests some open issues that help point out new research directions in HetVNETs.

We are very grateful to Prof. Xuemin (Sherman) Shen, the *SpringerBriefs* series editor on Wireless Communications. This book would not be possible without his kind support. Special thanks are also attributed to Jennifer Malat and Melissa Fearon at Springer Science+Business Media, for their assistance throughout the preparation process of this monograph.

We would like to thank Qiang Zheng, Haojun Yang, Lu Hou, Fei Liu, Xuemei Xin, and Zhiwei Zeng from the Wireless Signal Processing and Network (WSPN) group at the Beijing University of Posts and Telecommunications (BUPT) for their contributions to the work presented in this monograph. We also would like to thank all the members of the WSPN group for their valuable discussions and insightful suggestions, ideas, and comments.

This work is funded in part by the National Science Foundation of China (No. 61331009), National Key Technology R&D Program of China (No. 2015ZX03002009-004), and Fundamental Research Funds for the Central Universities (No. 2014ZD03-02).

Beijing, China   Kan Zheng
Beijing, China   Lin Zhang
Cairns, QLD, Australia   Wei Xiang
Beijing, China   Wenbo Wang

Contents

Acronyms

3D	Three-dimensional
ABF	Adaptive broadcast frame
AC	Access category
ACK	Acknowledgement
AF	Amplify-and-forward
AIFS	Arbitration interframe space
AMC	Adaptive modulation and coding
AWGN	Additive white Gaussian noise
BS	Base station
BSSID	Basic service set identification
CAM	Cooperative awareness message
CAP	Contention access period
CCH	Control channel
CCHI	CCH interval
CDF	Cumulative distribution function
CELL-DCH	CELL dedicated channel
CELL-FACH	CELL forward access channel
CELL-PCH	CELL paging channel
CH	Cluster head
CN	Core network
CQI	Channel quality indicator
CRP	Contention-based reservation period
CSMA	Carrier sense multiple access
CTS	Clear-to-send
CW	Contention window
D2D	Device-to-device
DEN	Decentralized environmental notification
DF	Decode-and-forward
DOT	Department of Transportation
DS-CDMA	Direct sequence code division multiple access
DSRC	Dedicated short-range communications

eMBMS	Evolved multimedia broadcast and multicast service
eNB	Evolved nodeB
EAP	Exclusive access period
EAW	Exclusive access window
ECA	Exclusive channel access
EDCA	Enhanced distributed channel access
EDCAF	Enhanced distributed channel access function
ETSI	European Telecommunications Standards Institute
FCFS	First come first serve
GI	Guard interval
GPS	Global Positioning System
GW	Gateway
HetVNET	Heterogeneous Vehicular NETwork
HLL	Heterogeneous link layer
I2V	Infrastructure-to-vehicle
ICI	Inter-carrier interference
IEEE	Institute of Electrical and Electronics Engineers
IPv4	Internet Protocol version 4
IPv6	Internet Protocol version 6
ISO	International Standards Organization
ITS	Intelligent transportation system
LBT	Listen-before-talk
LOS	Line-of-sight
LTE	Long-term evolution
MAC	Medium access control
MBMS	Multimedia Broadcast and Multicast Services
MBSFN	MBMS single frequency network
MCS	Modulation and coding scheme
MIMO	Multiple input multiple output
MSR	Maximum sum rate
NP	Nondeterministic polynomial
OBU	On-board unit
OFDM	Orthogonal frequency division multiplexing
OVSF	Orthogonal variable spreading factor
PCF	Point coordination function
PECA	Prioritized exclusive channel access
PHY	Physical
PS	Processor sharing
QCI	QoS class identifier
QoS	Quality of service
QPSK	Quadrature phase shift keying
RAN	Radio access network
RB	Resource block
RR	Round robin
RRC	Radio resource control

RSSI	Received signal strength indicator
RSU	Roadside unit
RTS	Request-to-send
SC	Service center
SCH	Service channel
SCHI	SCH interval
SF	Spreading factor
SNR	Signal-to-noise radio
TCP	Transmission Control Protocol
TDMA	Time-division multiple-access
UDP	User Datagram Protocol
UE	User equipment
URA-PCH	URA paging channel
UTC	Universal Coordinated Time
V2I	Vehicle-to-infrastructure
V2V	Vehicle-to-vehicle
VANET	Vehicular Ad hoc NETwork
VC	Vehicular cloud
VCC	Vehicular cloud computing
VE	Vehicle equipment
VoIP	Voice over Internet Protocol
VRRA	Virtual radio resource allocation
WAVE	Wireless access in vehicular environments
WBSS	WAVE basic service set
WCDMA	Wideband code division multiple access
WHO	World Health Organization
WLAN	Wireless local area network
WSA	WAVE service advertisement
WSMP	WAVE Short Message Protocol

Chapter 1
Introduction

The Heterogeneous Vehicular NETwork (HetVNET) is an important and timely topic, which is largely motivated by the fact that each type of current wireless networks not only offers its unique benefits, but also has its own drawbacks. Therefore, different from existing studies in the literature, this book focuses specifically on various services that can be provisioned by HetVNETs. In this chapter, we first present the motivations of the HetVNET after a short overview on the existing vehicular networks. Then, an analysis on the service and user case for the Intelligent transportation system (ITS) is given. The aim of the monograph is provided finally.

1.1 Motivation of Heterogeneous Vehicular Networks

In recent years, traffic congestion and accidents, alongside environmental pollution caused by road traffic and fuel consumption, have become compelling global issues. Both developing and developed countries are plagued by traffic problems. High traffic accident rates claim huge losses of life and property. As reported by the World Health Organization (WHO), more than 100 million people die in traffic accidents worldwide, and the resultant economic losses are up to $500 billion each year [1]. Moreover, traffic congestion in urban areas decreases the efficiency of transportation, and hence hinders economic growth. As a result, both the safety and the efficiency of current transportation systems have room for significant improvements.

Vehicular networks are designed to provide information exchange via Vehicle-to-Vehicle (V2V) and Vehicle-to-Infrastructure (V2I) communications. It is reported that over 50 % of interviewed consumers are highly interested in the idea of connected cars, 22 % of whom are willing to pay $30~65 per month for value-added on-road connectivity services [2]. In 1999, the Federal Communications

© The Author(s) 2016
K. Zheng et al., *Heterogeneous Vehicular Networks*, SpringerBriefs in Electrical and Computer Engineering, DOI 10.1007/978-3-319-25622-1_1

Commission (FCC) allocated 75 MHz (from 5.850 to 5.925 GHz) bandwidth to Dedicated Short Range Communications (DSRC) in vehicular environments. The US Department of Transportation (DOT) estimates that DSRC-based V2V communications can reduce up to 82 % of all road crashes in the USA, potentially saving thousands of lives and billions of dollars [3]. Meanwhile, much attention has been paid to the applicability of mobile cellular networks to support vehicular services, which can provide wide coverage and high data rate services to vehicular users. However, both DSRC and mobile cellular networks have their respective limitations when used in vehicular environments.

1.1.1   DSRC

DSRC is a wide-consensus wireless technology that is designed to support ITS applications in vehicular networks. In 1999, the US FCC allocated 75 MHz of licensed spectrum in the 5.9 GHz band to DSRC. The US DOT estimates that V2V communications based on DSRC can eliminate up to 82 % of all crashes involving unimpaired drivers in the USA, and about 40 % of all crashes occurred at intersections [3]. These statistics imply the huge potential for DSRC technology to reduce crashes and to improve safety for the driving public.

In February 2014, US Transportation Secretary Anthony Fox announced that the National Highway Traffic Safety Administration (NHTSA) would begin a V2V rule-making process and ultimately planned to require the life-saving communications technology to be installed in all new cars and light trucks. In August 2014, the National Highway Traffic Safety Administration issued an Advanced Notice of Proposed Rule Making (ANPRM), which suggests the US DOT may mandate DSRC in vehicles. To ensure traffic safety and deficiency, governments, academia, and auto-makers have been actively pursuing research on DSRC. The on-going US DOT Connected Vehicle Safety Pilot Program [4] conducts tests and collects data on the readiness and effectiveness of DSRC-based V2V and V2I communications for supporting collision prevention applications.

DSRC, which adopts IEEE 802.11p as its PHYsical (PHY) and Medium access control (MAC) layers, is derived from IEEE 802.11e with small modifications in the Quality of service (QoS) aspects. Easy deployment, low costs, and the capability to support V2V communications with the ad-hoc mode are its advantages compared to Long-term evolution (LTE) networks. However, DSRC has it own drawbacks, e.g., scalability issues, unbounded delays, limited radio range, short-lived V2I connectivity, and so on. As concluded in [2], even though the current DSRC technology is shown to be effective in supporting vehicular safety applications in numerous field trials, and to hold the promise for significantly reducing crashes, significant challenges remain for employing DSRC technology in some hostile vehicular environments.

1.1.2 Cellular Networks for Vehicular Communications

Although DSRC is widely considered as the *de facto* technique for vehicular networks, stakeholders have recently begun investigating the applicability of LTE to ITS services [2]. This is due to the following reasons: (1) large coverage; (2) high capacity; (3) centralized architecture; (4) high market penetration; and (5) multicast/broadcast support.

However, some issues remain to be resolved before cellular networks can be put to widespread use in vehicular communications. Firstly, the MAC layer lacks efficient scheduling mechanisms for a proper mapping of vehicular traffic features to the existing QoS class identifier (QCI) and/or new QCI definition. Secondly, when Multimedia Broadcast and Multicast Services (MBMS) are employed to broadcast vehicular service messages, the signaling overhead resulting from the subscription and joining procedures to a multicast service is overly large. Thus, it is essential to design lightweight joining/leaving procedures for dynamic groups of vehicles. Meanwhile, traditional applications offered by cellular networks, such as Voice over Internet Protocol (VoIP) and file sharing, may suffer from different levels of impact attributed to the new V2I traffic. Finally, effective business models are to be studied to support the widespread use of cellular networks for ITS applications.

1.1.3 Motivation of Heterogeneous Vehicular Networks

Based on the above discussions, a single wireless technology cannot well support ITS services, especially in dense traffic areas and heavy load environments. Therefore, the future trend of vehicular networks is to depart from focusing on a single technology towards designing systems built on multiple technologies. The European Telecommunications Standards Institute (ETSI), the International Standards Organization (ISO), and the US DOT are currently investigating the complementary roles of IEEE 802.11p, LTE, and other cellular technologies in supporting ITS applications cooperatively.

One of the key motivations for considering such heterogeneous vehicular networks is the widespread availability of these technologies, e.g., LTE, DSRC, and so on. Upon wide deployment, LTE is expected to play a crucial role in complementing the drawbacks of DSRC technology.

1.2 User Cases and Requirements for Safety and Non-Safety Related Services

The main objective of this section is to summarize user cases and services for the deployment of the ITS. The system capabilities are then derived from a requirement analysis of these user cases and services. ITS services can be broadly categorized into safety and non-safety services as described in [2, 5]. The former disseminates real-time safety-related messages, e.g., various warning messages including abrupt brake warning messages, so as to prevent car accidents, while the latter is to optimize the flow of vehicles in an effort to reduce the travel time and to improve the road users' experience.

1.2.1 Safety-Related Services and User Cases

Safety services aim at reducing the risk of car accidents and decreasing the possibility of life losses for vehicular users. Timeliness and reliability are considered as highly demanding requirements for this kind of services. Table 1.1 lists the requirements for the user cases of safety services [6, 7]. The safety-related services can be classified into four categories, i.e., Category I (vehicle status warning), Category II (vehicle type warning), Category III (traffic hazard warning), and Category IV (dynamic vehicle warning). The minimum frequency of periodic messages of a safety service varies from 1 to 10 Hz, and the reaction time of most drivers ranges from 0.6 to 1.4 s [8]. Thus, it is reasonable to restrict the maximum latency time to be no more than 100 ms. For example, the maximum latency of pre-crash warning is 50 ms. The communications mode and the user cases are also compared in Table 1.1. The security and reliability requirements are very strict due to the characteristics of safety services. We mainly consider two message types used for safety services [5], i.e.,

- **Cooperative awareness message (CAM)**: CAMs are sent or broadcast periodically to areas of interest primarily for road warning purposes, e.g., the user cases in Category II as shown in Table 1.1. The exchanged messages usually include the information on a vehicle's status, type, positions, speed, and so on.
- **Decentralized environmental notification (DEN)**: DEN messages are usually triggered by special events, e.g., the user cases in Categories I, III, and IV listed in Table 1.1. The purpose of DEN is to notify the vehicles in areas of interest of potential hazards.

1.2.2 Non-Safety Related Services and User Cases

Non-safety services are used primarily for traffic management, congestion control, improvement of traffic fluidity, infotainment, etc. The main objective of non-safety services is to offer a more efficient and comfortable driving experience. These services have no stringent requirements on latency and reliability [7]. As shown in Table 1.2, non-safety services can be roughly classified into two categories, i.e., Category I (traffic management) and Category II (infotainment). The former is to improve the traffic fluidity as well as offering secondary benefits not directly associated with traffic management [6]. For instance, due to efficient

Table 1.1 Safety services and user cases requirements

Category	User cases	Communication mode	Security reliability requirements	Minimum frequency of periodic messages	Maximum latency
Category I	Emergency electronic brake lights	Time limited periodic broadcast on event	High/High	10 Hz	100 ms
	Abnormal condition warning	Time limited periodic broadcast on event	High/High	1 Hz	100 ms
Category II	Emergency vehicle warning	Periodically triggered by vehicle mode	High/High	10 Hz	100 ms
	Slow vehicle warning	Periodically triggered by vehicle mode	High/High	2 Hz	100 ms
	Motorcycle warning	V2X cooperative awareness	High/High	2 Hz	100 ms
	Vulnerable road user Warning	V2X cooperative awareness	High/High	1 Hz	100 ms
Category III	Wrong way driving warning	Time limited periodic broadcasting on event	High/High	10 Hz	100 ms
	Stationary vehicle warning	Time limited periodic broadcasting on event	High/High	10 Hz	100 ms
	Traffic condition warning	Time limited periodic messages broadcast-ing/authoritative message triggered	High/High	1 Hz	100 ms
	Signal violation warning	Temporary messages broadcasting on event	High/High	10 Hz	100 ms
	Roadwork warning	Temporary messages broadcasting on event	High/High	2 Hz	100 ms
	Decentralized floating car data	Time limited periodic broadcasting on event	High/High	10 Hz	100 ms
Category IV	Overtaking vehicle warning	V2X cooperative awareness	High/High	10 Hz	100 ms
	Lane change assistance	V2X cooperative awareness	High/High	10 Hz	100 ms
	Pre-crash sensing warning	Broadcast of pre-crash state	High/High	10 Hz	50 ms
	Cooperative glare reduction	V2X cooperative awareness	Medium/Medium	2 Hz	100 ms

Table 1.2 Non-safety service and user cases requirements

Category	User cases	Communication mode	Security reliability requirements	Minimum frequency of periodic messages	Maximum latency
Category I	Regulatory/contextual speed limits	Authoritative message triggered by traffic management entity	High/High	1 Hz	N/A
	Traffic light optimal speed advisory	Periodic, permanent messages broadcasting	High/High	2 Hz	100 ms
	Intersection management	Periodic, permanent messages broadcasting	High/High	1 Hz	100 ms
	Cooperative flexible lane change	Periodic messages broadcasting	High/High	1 Hz	500 ms
	Electronic toll collect	I2V broadcasting and unicast full duplex session	High/High	1 Hz	500 ms
Category II	Point of interest notification	Periodic, permanent messages broadcasting	Medium/Medium	1 Hz	500 ms
	Local electronic commerce	Duplex communication between RSU and vehicles	High/High	1 Hz	500 ms
	Media download	User access to Internet for multimedia download	Medium/Medium	1 Hz	500 ms
	Map download and update	Access to Internet for map download and update	Medium/Medium	1 Hz	500 ms

traffic scheduling, the travel time and fuel consumption can be reduced. The latter provides on-demand entertainment information to passing vehicles. Compared to safety services, non-safety services have varying QoS requirements. For most non-safety services, the minimum frequency of periodic messages is 1 Hz, while the maximum latency is 500 ms. In special user cases such as optimal traffic light speed advisory and intersection management, positioning accuracy is crucial, e.g., no more than 5 m [6]. On the other hand, user cases such as local electronic commerce with monetary transactions require high-level security. For traffic efficiency services, broadcast messages have to be authoritative and endorsed by traffic management authorities.

1.3 Aim of the Monograph

Several wireless communication systems have been considered for supporting ITS services via V2V and V2I communications. Among them, DSRC and cellular networks such as LTE technology are front runners, and both are considered well suited for providing ITS services under the condition of low vehicle density [9, 10]. However, with an ever increasing number of vehicles, LTE networks can be easily overloaded. Moreover, the work in [9] shows that DSRC in conjunction with IEEE 802.11p exhibits poor performance in the event of a large number of vehicles. As a result, to remedy the drawbacks of existing vehicular networks, the HetVNET is needed to support various services under dense vehicular environments.

The aim of this monograph is to investigate how to utilize all the resources in an HetVNET taking into account different scenarios and applications. Specifically, we focus on the following research directions, i.e., (1) how to coherently integrate all the wireless communications systems in HetVNETs; (2) how to manage and maintain communications among the vehicles and between the vehicles and the infrastructure; and (3) how to make efficient use of all the radio resources in the HetVNET. To address these questions, in this monograph, we analyze the network architecture of the HetVNET, and provide an in-depth study into the MAC mechanisms for vehicular networks as well as cooperative scheduling schemes. The results from these studies provide an insight into and the guidelines on the design and deployment of future HetVNETs.

References

[1] F. Martinez, C.-K. Toh, J.-C. Cano, C. Calafate, and P. Manzoni, "Emergency services in future intelligent transportation systems based on vehicular communication networks," *IEEE Trans. Intell. Transp. Syst. Mag.*, vol. 2, no. 2, pp. 6–20, Oct. 2010.
[2] G. Araniti, C. Campolo, M. Condoluci, A. Iera, and A. Molinaro,, "LTE for vehicular networking: A survey," *IEEE Commun. Mag.*, vol. 51, no. 5, pp. 148–157, May 2013.

[3] J. Kenney, "Dedicated short-range communications (DSRC) standards in the united states," *Proceedings of the IEEE*, vol. 99, no. 7, pp. 1162–1182, Jul. 2011.

[4] X. Wu, S. Subramanian, R. Guha, R. White, J. Li, K. Lu, A. Bucceri, and T. Zhang, "Vehicular communications using DSRC: Challenges, enhancements, and evolution," *IEEE J. Sel. Areas Commun.*, vol. 31, no. 9, pp. 399–408, Jul. 2013.

[5] "Intelligent transport systems (ITS); framework for public mobile networks in cooperative ITS (C-ITS)," European Telecommunications Standards Institute (ETSI), Tech. Rep. 102 962 V1.1.1, Feb. 2012.

[6] "Intelligent transport system (ITS); vehicular communications; basic set of applications; definitions," European Telecommunications Standards Institute (ETSI), Tech. Rep. 102 638 V1.1.1, Jun. 2009.

[7] "Vehicle safety applications," US DOT IntelliDrive(SM) Project-ITS Joint Program Office, Tech. Rep., 2008.

[8] Y. Wu, H. Yuan, H. Chen, and J. Li, "A study on reaction time distribution of group drivers at car-following," in *Proc. Second International Conference on Intelligent Computation Technology and Automation (ICICTA)*, Changsha, Hunan, Oct. 2009, pp. 452–455.

[9] C. Han, M. Dianati, R. Tafazolli, R. Kernchen, and X. Shen, "Analytical study of the IEEE 802.11p MAC sublayer in vehicular networks," *IEEE Trans. Intell. Transp. Syst.*, vol. 13, no. 2, pp. 873–886, Feb. 2012.

[10] M. Kihl, K. Bur, P. Mahanta, and E. Coelingh, "3GPP LTE downlink scheduling strategies in vehicle-to-infrastructure communications for traffic safety applications," in *Proc. IEEE Symposium on Computers and Communications (ISCC)*, Cappadocia, Turkey, Jul. 2012, pp. 448–453.

Chapter 2
Architecture of Heterogeneous Vehicular Networks

2.1 Background

Due to the high mobility of vehicles and the dynamic topology changes of Vehicular Ad hoc NETwork (VANET), it is difficult to provide satisfied ITS services only through a single wireless network. Consequently, by integrating different wireless access networks such as LTE and DSRC, the HetVNET is expected to be a good platform that can meet various demanding communications requirements of ITS services. In this chapter, we first present a framework of the HetVNET [1]. Several HetVNET candidate communications techniques are then discussed for comparison purposes.

2.2 Framework of Heterogeneous Vehicular Networks

As illustrated in Fig. 2.1, an HetVNET is composed of three main components, namely a Radio access network (RAN), a Core network (CN), and a Service center (SC). Service providers can often supply a variety of services to vehicular users through the SC. The CN is a key component of the HetVNET, because it provides many important functions, such as aggregation, authentication, and switching. In this book, we focus only on the RAN. In the HetVNET, there are two types of communications links, i.e., V2V and V2I, which are similar to traditional vehicular networks supported by only a single communications technology [2–4]. V2V allows for short- and medium-range communications among vehicular users, offering low deployment costs and supporting low-latency message delivery. V2I enables vehicles to connect to the Internet for information dissemination and infotainment via a roadside Base station (BS). Various candidate wireless access technologies can be used to support V2I and V2V communications subject to specific requirements.

© The Author(s) 2016
K. Zheng et al., *Heterogeneous Vehicular Networks*, SpringerBriefs in Electrical and Computer Engineering, DOI 10.1007/978-3-319-25622-1_2

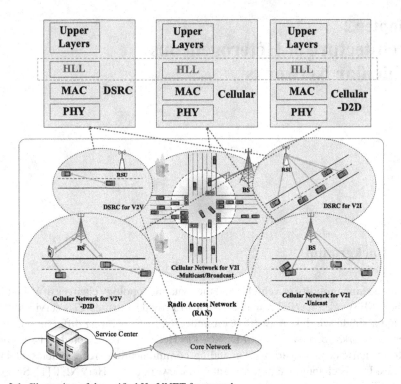

Fig. 2.1 Illustration of the unified HetVNET framework

Thus, it is a challenging task to select an efficient and suitable radio access method that meets all distinct QoS requirements of desired services for vehicular users.

One of challenges in HetVNETs is to support a dynamic and instant composition of different networks, and to allow operators to utilize radio resources in an efficient and flexible manner. Towards this end, as illustrated in Fig. 2.1, we introduce a new layer, namely the heterogeneous link layer (HLL), which operates on the top of the MAC layer in each radio access network. The HLL enables unified processing, offers a unified interface to the higher layers, and can adapt to the underlying radio access techniques. In the proposed layer, we define specific functions for joint management of multi-radio resources and load sharing among different networks. These functions facilitate efficient link-layer inter-working among multiple networks. There exists a trade-off between the performance and exchange overhead of the HetVNET, which depends on the possible operational time scale of the HLL. The main objectives of the HLL functions are to enable the global management of network resources, and to meet the QoS requirements of safety/non-safety services by facilitating the coordination among various radio networks.

Since physical layer techniques and network layer protocols for different systems often have their own unique characteristics, a unified approach to enable cooperation among multiple systems is highly desirable. Through virtualization techniques, the physical wireless infrastructure and the radio resources in HetVNETs can be abstracted and isolated into a number of virtual resources, which are then shared by multiple parties through isolating each other [5]. Thus, our purpose is to introduce virtualization functions to the HLL for abstracting, slicing, isolating, and sharing resources so that each wireless system in the HetVNET can be regarded as part of the entire network. However, the unique characteristics of different wireless systems, in terms of physical radio resources, Medium access control (MAC), and network protocols, etc., make this task extremely complicated.

Virtualization can be implemented at different levels, ranging from the spectrum level through to the physical radio resource unit, which determines the flexibility of radio resource utilization. Virtualization at a higher level may reduce the flexibility of virtualization, while better multiplexing of resources across slices results in a more feasible implementation. However, this may lead to less efficient use of resources and less strict isolation between different systems. For example, in spectrum-level slicing, resource sharing between the LTE and DSRC systems emphasizes on the data bearer instead of the physical layer technique. A vehicular user may be restricted to either LTE or DSRC through the access control function at the HLL with the knowledge of traffic loading and resource usage of different systems. On the other hand, when wireless virtualization is implemented at a lower level with a different definition of slices, the effect may be opposite. It is possible that the physical resources that belong to one or more wireless systems are virtualized and split into virtual resource slices [6], which can be either bandwidth-based or resource-based [7]. Then, virtual radio resource allocation (VRRA) can be implemented to map between the physical and virtual links, allowing for dynamic allocation of radio resources to different systems [8].

2.3 V2I Communications

V2I communications aim to connect vehicles to the infrastructure located on the roadside. Since the infrastructure of cellular networks has been widely deployed in the past decades, it is economical to utilize cellular networks to support V2I communications [9, 10]. Another solution is to use DSRC, which is based on the IEEE 802.11p/1609 wireless access in vehicular environments (WAVE) protocols [11].

2.3.1 Cellular-Network-Based V2I Communications

Cellular networks offer two transmission modes for V2I communications, namely unicast and multicast/broadcast. Unicast can be used for both uplink and downlink message distributions, which is point-to-point communications between a vehicle and the BS, also known as the Evolved nodeB (eNB). On the other hand, multicast/broadcast is exclusively used for the distribution of downlink messages, which

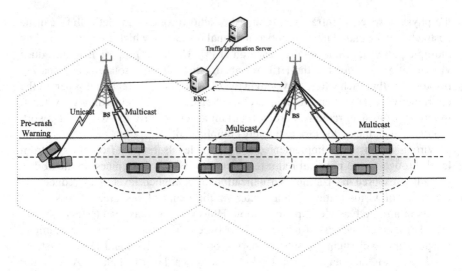

Fig. 2.2 Example of multicast/broadcast by BSs in the HetVNET

refers to point-to-multi-point transmission. In the broadcast scenario illustrated in Fig. 2.2, the traffic information server may distribute the safety messages of "pre-crash warning" to different broadcast areas via MBMS [12]. Each broadcast area consists of multiple cells configured by mobile operators.

2.3.1.1   WCDMA for V2I

Wideband code-division multiple access (WCDMA) is one of the most successful cellular systems. It is based on the direct sequence code-division multiple access (DS-CDMA) technique, where the signal of a physical channel is spread over wide bandwidth through multiplying with a certain channelization code, e.g., the orthogonal variable spreading factor (OVSF) code. This unique code distinguishes between each physical channel in a WCDMA system. The radio resource control (RRC) defines protocols that describe which processes should be active in a vehicle equipment (VE), and whether a common or dedicated/shared channel is used [12]. In accordance with the inactive and active statues of VEs, the sub-states can be classified as RRC idle and RRC connected, respectively, as illustrated in Fig. 2.3. Typically, an inactive VE stays in the RRC Idle state, which is a power saving state with little signaling traffic. As for the RRC connected state, there are four different sub-states, i.e., the CELL dedicated channel (CELL-DCH), CELL forward access channel (CELL-FACH), CELL paging channel (CELL-PCH), and URA paging channel (URA-PCH). When a dedicated channel is allocated to a VE, i.e., in the CELL_DCH, messages can be transmitted and received with minimal latency. After a certain inactive period A (usually 2 s), the VE transits to the CELL_FACH, which

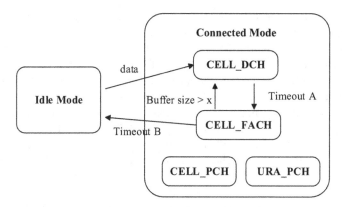

Fig. 2.3 Illustration of the WCDMA RRC states for V2I communications

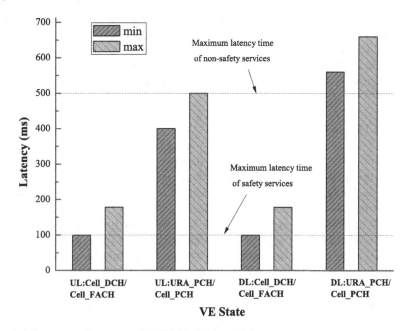

Fig. 2.4 Latency performances of WCDMA different RRC states

can be used to exchange control information and a small amounts of user data. When the buffer in the VE or RNC exceeds a certain threshold (i.e., about 220 bytes on the uplink), the VE sends an RRC measurement report, and thus initiates a channel type switch to the CELL_DCH. In the CELL_PCH, the VE sends regular cell updates and thus becomes known at the cell level. The URA_PCH state is similar to that of the CELL_PCH, except that the VE sends URA updates instead of cell updates.

However, there are still a few technical challenges that remain to be solved, when applying WCDMA to V2I communications. Figure 2.4 shows the delivery latency of the VE under different states in the WCDMA system. In the idle state, the connection setup requires $2 \sim 2.5$ s, which is not presented in Fig. 2.4. As can be seen from the figure, the delivery latency that the VE is in all the states is longer than the allowed maximum latency for safety services, i.e., 100 ms. This means that the WCDMA system cannot well support safety services in vehicular communications. On the other hand, in most scenarios, the WCDMA system can nearly meet the latency requirement of non-safety services, i.e., no more than 500 ms. For example, if the VE is in the CELL_DCH or CELL_FACH state, the latency is around 100–178 ms. When the VE is in the URA_PCH state, the latency is increased to $400 \sim 500$ ms. The latency of unicast is similar to that of the above case. For VEs in the CELL_DCH or CELL_FACH state, the delay is similar to that of uplink transmission. For VEs in URA_PCH, channel switching requires 300 ms. Furthermore, the paging procedure takes another 160 ms. Providing that the VEs are permanent in the CELL_DCH or CELL_FACH state, the delivery latency for most non-safety services may not be the bottleneck. In fact, the capacity of the WCDMA system is limited, in which a large number of VEs cannot always remain connected, i.e., in the CELL_DCH or CELL_FACH sub-state. Therefore, the number of VEs that can be in CELL_DCH and CELL_FACH simultaneously becomes a major limiting factor for non-safety services in the WCDMA system.

2.3.1.2   LTE for V2I

As stated in [13], LTE can provide uplink data rates up to 50 Mbps, and downlink data rates up to 100 Mbps with a bandwidth of 20 MHz, and supports a maximum mobile speed of 350 km/h The flat architecture of the LTE system is attributed to the low transmission latency, e.g., the theoretical round-trip time is lower than 10 ms, and the transmission latency in the RAN is up to 100 ms [14]. Therefore, LTE is envisioned to well support V2I communications. Especially, in the initial deployment stage of vehicular networks, LTE is expected to play a crucial role in supporting vehicular services. This could first take place in rural areas, where the vehicle density is low.

In general, LTE networks are capable of providing high capacity with wide coverage. For instance, LTE can support up to 1200 vehicles per cell in rural environments with a uplink delay under 55 ms and one CAM per second [12]. Besides, it also can provide a robust mechanism for mobility management. Experiments of trialing LTE in vehicles to support various applications, e.g., infotainment, diagnostics, and navigation, have been carried out. The results show that the LTE system is able to provide a data rate of 10 Mbps with a speed up to 140 km/h [15]. LTE can be particularly helpful at intersections by enabling a reliable exchange of cross-traffic assistance applications [14]. In [16], the authors analyze the applicability of LTE to vehicular safety communications at intersections. Their analysis shows that the LTE system can support the demand of transmitting approximate 1500 CAMs per second per cell. Furthermore, the Evolved multimedia broadcast and multicast service (eMBMS) is an effective means to support multicast or broadcast services in highly dense vehicle environments.

Nevertheless, several problems need to be solved before LTE systems can be widely used for V2I communications [14]. Firstly, the MAC layer of LTE lacks an efficient scheduling mechanism for properly mapping the vehicular traffic features to the existing QCI and/or the new QCI definition. Secondly, when the eMBMS is employed to broadcast vehicular service messages, the signaling overhead resulting from the subscribing and joining procedures to the multicast service is overly large. Thus, it is essential to design lightweight joining/leaving procedures for dynamic groups of vehicles. The challenge is how to ensure transmission efficiency while reducing the overhead. Meanwhile, traditional applications offered by LTE networks may be affected by different levels of potential impact due to the introduction of new types of traffic, especially the heavy load ones [17].

2.3.2 DSRC-Based V2I Communications

Figure 2.5 illustrates the WAVE protocol in a DSRC network. In order to enable robust connections and fast setup for moving vehicles, the half-clocked mode with a 10 MHz bandwidth in the physical layer, termed IEEE 802.11p, is employed. Considering the characteristics of vehicular environments, the Enhanced distributed channel access (EDCA) mechanism in IEEE 802.11e with small modifications is adopted to satisfy the strict QoS requirements of the MAC layer [18]. In order to meet the requirements of vehicular communications, a suite of standards are defined by the IEEE 1609 Working Group for DSRC networks, i.e., 1609.4 for Channel Switching, 1609.3 for Network Services including the WAVE Short Message Protocol (WSMP), and 1609.2 for Security Services. In order to avoid the packetization overhead, the minimum WSM overhead is 5 bytes, and even with optional extensions the overhead rarely exceeds 20 bytes. The minimum overhead associated with User Datagram Protocol (UDP)/Internet Protocol version 6 (IPv6) is 52 bytes. In addition, in the network and transport layers, the Internet Protocol version 4 (IPv4), Transmission Control Protocol (TCP), and UDP are also employed on the top of the stack. The SAE J2735 Message Set Dictionary standard specifies a set of message formats that support a variety of vehicle-based applications [11]. DSRC networks can operate well under sparse nomadic deployment with stationary channels. However, vehicular communications may take place over severe frequency-selective multipath and fast fading channels, as well as in densely populated environments. Therefore, there is a large room for improvement and enhancement in DSRC. Next, several problems of DSRC networks when used for V2I communications are discussed.

- **Sparse pilot design**: The dynamic V2I environment with large multipath delay spread and high mobility results in highly time-frequency selective vehicular communications channels. In a typical application scenario, 50 % coherence

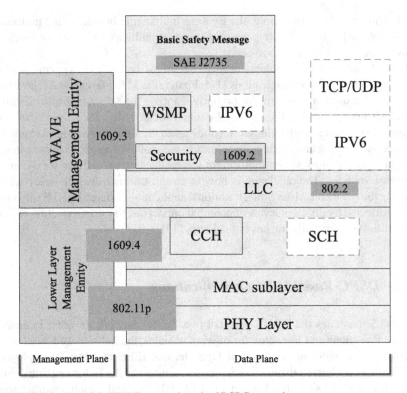

Fig. 2.5 Illustration of the WAVE protocol stack of DSRC networks

bandwidth is roughly in the order of 1 MHz, and 50 % of the coherence time can be as short as 0.2 ms [19]. Then, a typical packet transmission period in DSRC, i.e., approximately 0.5 ms with a packet size of 300 bytes, Quadrature phase shift keying (QPSK) modulation, and a code rate of 1/2, is larger than the coherence time. Moreover, the inter-spacing between two pilot subcarriers defined in IEEE 802.11p, i.e., 2.4 MHz, is larger than the coherence bandwidth. Thus, such a sparse pilot design is insufficient to accurately estimate the channel state information. The only way is to improve the receiver performance at the expense of implementation complexity;

- **Channel congestion**: When the Carrier sense multiple access (CSMA) mechanism is employed at the MAC layer of the DSRC network, the probability of collisions increases rapidly with the number of vehicles in the network, resulting in large end-to-end latency and low channel utilization [20]. Therefore, channel congestion has to be dealt with so as to guarantee the QoS requirements of vehicular services. One approach is to reduce the number of transmitters to within the carrier sense range of each device [21, 22];
- **Unbalanced link**: Due to the different hardware configurations between the Onboard Unit (OBU) in the vehicle and the Roadside Unit (RSU), the coverage

areas of the OBU and RSU are obviously different, causing the so-called unbalanced link problem. For example, the reliable radio communications range from the RSU to OBU is up to 1,100 m, while the range from the OBU to RSU is only up to 400 m. Thus, the OBU may commence data transmission after moving into the broadcast range of an RSU, even at a distance that is too far for the RSU to receive data from the OBU [23, 24]. Then, communications quality deteriorates due to such "unbalanced links"; and

- **Prioritization and service selection**: This situation only arises in the overlapped coverage area of multiple RSUs. When an OBU moves into such an overlapped area, various services are provided by multiple RSUs. The OBU may create a WAVE basic service set (WBSS) with the first RSU it detects. It may switch to another RSU only if that RSU is advertising a service with a higher priority. If the services from the other RSUs have lower priorities compared with the first one, the OBU does not create a WBSS with any other RSUs, and may miss any service channel messages or services offered by the other RSUs [23]. The wildcard WBSS is an efficient method to resolve this problem. In the event of overlapped coverage, an RSU can configure its basic service set identification (BSSID) with wildcard BSSID, i.e., 0xFFFFFF, so that the OBUs already in a WBSS can still receive frames, and do not miss any services offered by the other RSUs which use the wildcard BSSID [11].

2.4 V2V Communications

V2V communications refer to direct connection between vehicles. It aims at minimizing traffic accidents and improving traffic efficiency. Accidents caused by slow vehicles or non-sight vehicles may be avoided by exchanging information on velocity, acceleration, and vehicle status with neighboring vehicles. Extensive investigations and trials on V2V have been carried out with the objective of supporting traffic services, such as slow vehicle warning and abnormal vehicle status warning [25]. In this subsection, two candidate techniques for V2V communications are discussed in detail.

2.4.1 LTE D2D-Based V2V Communications

Device-to-device (D2D) communications underlaying a cellular network have been proposed as a means of taking advantage of the physical proximity of communicating devices in LTE systems [26, 27]. In the D2D mode, user equipments (UE) in close proximity can directly communicate with each other. As a candidate technique supporting V2V in HetVNETs, D2D communications in LTE face several challenges. Since D2D communications links share the same radio resources with other links in the LTE network, interference is a major issue when employing

D2D in HetVNETs. For example, in the FDD system, when a D2D link uses downlink resources, the donor eNB may cause severe interference to the D2D pair. Moreover, the interference from neighboring cells is another problem facing D2D communications. On the other hand, if a D2D pair uses uplink resources, the receiving end of the D2D pair may suffer strong interference from a cellular UE using the same uplink resources.

Most D2D devices in LTE systems are usually static or of low-speed mobility. However, vehicles usually move in medium or high speeds, which may severely degrade the performance of D2D communications. Specifically, existing peer and service discovery of D2D communications does not work well in vehicular environments. In the D2D mode, before any two vehicles can directly communicate with each other, they need to first discover the existence of its peer, which is a time-consuming procedure. As specified in [28], the discovery period usually is set to 1, 2, 5, or 10 s. Since the survival time of available connectivity between two vehicles is very short in vehicular environments, it is very difficult for the existing D2D discovery mechanism to meet the strict QoS requirements of safety services. Taking as an example the safety user case of hard-braking warning, we assume that two vehicles move at a speed of 120 km/h (i.e., 33.3 m/s) along the same direction with an inter-vehicle spacing of 30 m. If the front vehicle starts hard-braking with a deceleration of $4 \, m/s^2$ and the reaction time of the rear vehicle's driver is about 1.5 s [29], the time remained for message transmission is only around 3 s. Thus, in many cases, the D2D discovery time is larger than that allocated for message transmission, which is not acceptable for delivering safety messages with strict QoS requirements.

2.4.2   DSRC-Based V2V Communications

DSRC has been shown to be effective in supporting both safety and non-safety services in V2V communications. Firstly, V2V communications usually employ a decentralized approach, in which the network is autonomous and needs no external infrastructure to organize itself. Secondly, since both entities in V2V communications are vehicles, there is no aforementioned "unbalanced link" problem in V2I communications. Furthermore, V2V communications based on DSRC do not interfere with cellular networks due to the use of different frequency bands. However, there still exist several challenges for using DSRC in V2V communications [30–32]. For example, in a densely populated vehicular environment, collisions occur so frequently attributed to the limitation of the CSMA mechanism that the overall performance significantly deteriorates.

2.5 Typical Application Scenarios

Each candidate technique for either V2I or V2V communications has its own advantages and disadvantages. For different application scenarios in HetVNETs, any candidate technique may be chosen according to their characteristics, and they can work together with the aid of the HLL. Two examples are given below to illustrate the applicability of HetVNETs to ITS services.

2.5.1 Urban Intersection Scenario

Figure 2.6 depicts a safety driving user case in an urban intersection scenario. Under this scenario, DSRC is used for the communications between vehicles, i.e., V2V communications, while LTE is employed to provide connections between the vehicles and the eNBs, i.e., V2I communications. The following cases (but not limited to) have to be considered for safety driving in such an urban intersection:

- **Collaboration between vehicle and eNB**: Pedestrians and obstacles are detected and reported to the eNB by vehicles or pedestrians. There are several methods to report roadwork, obstacles, and accidents to the eNB [33]. The traditional method

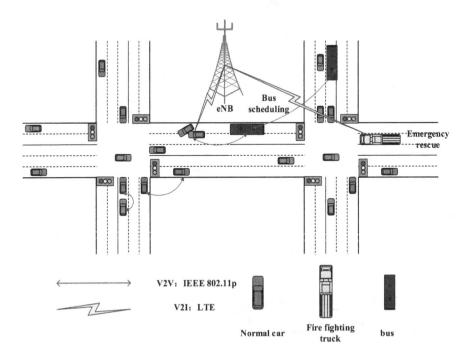

Fig. 2.6 Illustration of an urban intersection scenario

is that the witness sends the information to the eNB. A new method of notification may be like eCall [33], which is the most important road safety efforts made under the European Union's eSafety initiative. Based on the information (speed, direction, or target destination) that is periodically sent by vehicles, the eNB can predict mobility via some prediction algorithm, e.g., road-topology-based [34] and behavior-based mobility prediction [35]. Then, in order to avoid traffic congestion or accidents, the eNB can broadcast existing dead zones to the vehicles that may go through its coverage area;

- **Collaboration between vehicles**: The front vehicle is able to inform the following vehicles of sharp stops, and thus avoids the rear-end problem. Moreover, vehicles involved in a car accident may broadcast the occurrence of such an event so as to prevent further collisions; and
- **Traffic light management**: The duration of a traffic signal can be intelligently adjusted to pass high priority vehicles, such as fire-fighting trucks and buses.

2.5.2 Expressway Scenario

In the expressway scenario, there are generally two types of traffic flows, i.e., the free and synchronized flows as shown in Fig. 2.7a and b, respectively. These two vehicle flows may switch to each other.

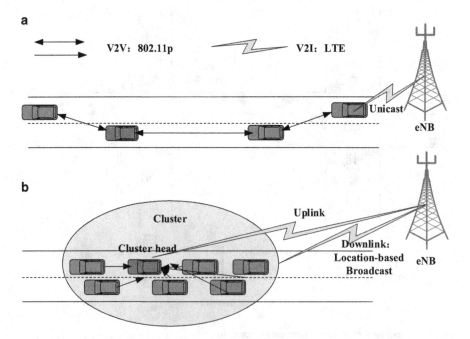

Fig. 2.7 Illustration of an expressway scenario. (**a**) Free flow. (**b**) Synchronized flow

- **Free flow**: In this flow, the number of vehicles in the HetVNET is small, and the interactions among vehicles are infrequent. Therefore, vehicles move with high speeds and the network topology changes rapidly so that the radio links are unreliable. In this case, the mobile cellular network such as the LTE system is preferred for V2I communications. However, in specific environments, e.g., in a tunnel, the received signal from the eNB is not of high quality at the vehicles. Then, the vehicles may help one another through multi-hop DSRC transmission before connecting to the eNB eventually [36]; and
- **Synchronized flow**: The traffic density of a synchronized flow is much higher, meaning that broadcast messages are likely to be flooded. Due to traffic jam, vehicle speeds are low, and the random behavior of vehicles can be modeled by a car-following behavior, meaning that the radio links among vehicles become relatively static. With the aid of DSRC, clustering mechanisms may be an efficient information dissemination method. The vehicles within the transmission range of DSRC form a cluster, and a Cluster head (CH) is elected via a certain algorithm. Then, on the V2I uplink, the CHs aggregate the data of their cluster members before forwarding it to the eNB via LTE. In this way, the overall LTE traffic can be reduced compared to separate transmissions by individual vehicular users [37]. For the downlink, the multicast of the LTE network can be used to distribute messages.

2.6 Summary

Since a single wireless communications network, either DSRC or LTE, cannot well satisfy the QoS requirements of ITS services, we propose an HetVNET framework. Several candidate techniques, e.g., DSRC and LTE cellular networks, are discussed and summarized in Table 2.1. As can be seen from the table, LTE is much more suitable for V2I communications than DSRC. On the contrary, DSRC is more practical for V2V communications than LTE D2D. The collaborations between heterogeneous networks are essential for HetVNETs.

Table 2.1 Advantages and challenges of candidate techniques for the HetVNET

Communications	LTE/LTE D2D	DSRC
V2I Communications	Advantages: • Large coverage • Robust mechanisms for mobility management • High downlink and uplink capacity • Centralized and flat architecture • High-efficiency eMBMS Challenges: • Lack of efficient scheduling schemes for ITS services • Users in the idle state cause high delays in disseminating messages • Prone to overloaded in high density environments	Advantages: • Easy deployment and low costs • Suitable for local message dissemination, i.e., traffic signal, parking information, etc. Challenges: • Sparse pilot design • Serious channel congestion with a large number of vehicles • Unbalanced link • Prioritization and service selection • Broadcast storm and hidden node problems
V2V Communications	Advantages: • High spectral efficiency • High energy efficiency • Efficient scheduling on D2D resources Challenges: • Interference between D2D pair and other users • Peer and service discovery is time consuming • Serious performance degradation in high vehicle speed	Advantages: • Easy deployment and low costs • Ad-hoc mode • Overhead of WSM messages is low Challenges: • Sparse pilot design • Serious channel congestion with a large number of vehicles • Adjacent band leakage in multi-channel operations • Broadcast storm and hidden node problems

References

[1] K. Zheng, Q. Zheng, P. Chatzimisios, W. Xiang, and Y. Zhou, "Heterogeneous vehicular networking: A survey on architecture, challenges and solutions," *IEEE Commun. Surveys Tuts.*, vol. PP, no. 99, pp. 1–1, 2015, DOI: 10.1109/COMST.2015.2440103.

[2] R. Atat, E. Yaacoub, M.-S. Alouini, and F. Filali, "Delay efficient cooperation in public safety vehicular networks using LTE and IEEE 802.11p," in *Proc. IEEE Consumer Communications and Networking Conference (CCNC)*, Las Vegas, NV, Jan. 2012, pp. 316–320.

[3] C. Ide, B. Dusza, M. Putzke, and C. Wietfeld, "Channel sensitive transmission scheme for V2I-based floating car data collection via LTE," in *Proc. IEEE International Conference on Communications (ICC)*, Ottawa, ON, Jun. 2012, pp. 7151–7156.

[4] E. Yaacoub and N. Zorba, "Enhanced connectivity in vehicular ad-hoc networks via V2V communications," in *Proc. International Wireless Communications and Mobile Computing Conference (IWCMC)*, Sardinia, Jul. 2013, pp. 1654–1659.

[5] C. Liang and F. Yu, "Wireless network virtualization: A survey, some research issues and challenges," *IEEE Commun. Surveys Tuts.*, vol. PP, no. 99, pp. 1–1, 2014.

[6] K. Pentikousis, Y. Wang, and W. Hu, "Mobileflow: Toward software-defined mobile networks," *IEEE Commun. Mag.*, vol. 51, no. 7, pp. 44–53, Jul. 2013.

[7] R. Kokku, R. Mahindra, H. Zhang, and S. Rangarajan, "NVS: A substrate for virtualizing wireless resources in cellular networks," *IEEE/ACM Trans. Netw.*, vol. 20, no. 5, pp. 1333–1346, Oct. 2012.

[8] L. Caeiro, F. D. Cardoso, and L. M. Correia, "Adaptive allocation of virtual radio resources over heterogeneous wireless networks," in *Proc. European Wireless Conference*, Poznan, Poland, Apr. 2012, pp. 1–7.

[9] M. Kihl, K. Bur, P. Mahanta, and E. Coelingh, "3GPP LTE downlink scheduling strategies in vehicle-to-infrastructure communications for traffic safety applications," in *Proc. IEEE Symposium on Computers and Communications (ISCC)*, Cappadocia, Turkey, Jul. 2012, pp. 448–453.

[10] S.-T. Cheng, G.-J. Horng, and C.-L. Chou, "Using cellular automata to form car society in vehicular ad hoc networks," *IEEE Trans. Intell. Transp. Syst.*, vol. 12, no. 4, pp. 1374–1384, Jun. 2011.

[11] J. Kenney, "Dedicated short-range communications (DSRC) standards in the united states," *Proceedings of the IEEE*, vol. 99, no. 7, pp. 1162–1182, Jul. 2011.

[12] "Intelligent transport systems (ITS); framework for public mobile networks in cooperative ITS (C-ITS)," European Telecommunications Standards Institute (ETSI), Tech. Rep. 102 962 V1.1.1, Feb. 2012.

[13] *Evolved Universal Terrestrial Radio Access (E-UTRA); LTE physical layer; General description*, 3GPP Std. 36.201, Rev. 12.2.0, Mar. 2015.

[14] G. Araniti, C. Campolo, M. Condoluci, A. Iera, and A. Molinaro,, "LTE for vehicular networking: A survey," *IEEE Commun. Mag.*, vol. 51, no. 5, pp. 148–157, May 2013.

[15] J. Mosyagin, "Using 4G wireless technology in the car," in *Proc. International Conference on Transparent Optical Networks (ICTON)*, Munich, Jun. 2010, pp. 1–4.

[16] T. Mangel, T. Kosch, and H. Hartenstein, "A comparison of UMTS and LTE for vehicular safety communication at intersections," in *Proc. IEEE Vehicular Networking Conference (VNC)*, Jersey City, Dec. 2010, pp. 293–300.

[17] K. Zheng, S. Ou, J. Alonso-Zarate, M. Dohler, F. Liu, and H. Zhu, "Challenges of massive access in highly dense LTE-advanced networks with machine-to-machine communications," *IEEE Wireless Commun.*, vol. 21, no. 3, pp. 12–18, Jun. 2014.

[18] "IEEE standard for information technology– local and metropolitan area networks– specific requirements– part 11: Wireless lan medium access control (MAC) and physical layer (PHY) specifications amendment 6: Wireless access in vehicular environments," *IEEE Std 802.11p-2010*, pp. 1–51, Jul. 2010.

[19] X. Wu, S. Subramanian, R. Guha, R. White, J. Li, K. Lu, A. Bucceri, and T. Zhang, "Vehicular communications using DSRC: Challenges, enhancements, and evolution," *IEEE J. Sel. Areas Commun.*, vol. 31, no. 9, pp. 399–408, Jul. 2013.

[20] C. Han, M. Dianati, R. Tafazolli, R. Kernchen, and X. Shen, "Analytical study of the IEEE 802.11p MAC sublayer in vehicular networks," *IEEE Trans. Intell. Transp. Syst.*, vol. 13, no. 2, pp. 873–886, Feb. 2012.

[21] C.-L. Huang, Y. P. Fallah, R. Sengupta, and H. Krishnan, "Adaptive intervehicle communication control for cooperative safety systems," *IEEE Netw.*, vol. 24, no. 1, pp. 6–13, Jan. 2010.

[22] A. Weinfeld, "Methods to reduce DSRC channel congestion and improve V2V communication reliability," in *Proc. 17th ITS World Congress*, Busan, Oct. 2010.

[23] S. Andrews and M. Cops, "Final report: Vehicle infrastructure integration proof of concept technical description-vehicle," VII Consortium, Tech. Rep., Feb. 2009.

[24] R. Kandarpa and M. Chenzaie, "Final report: Vehicle infrastructure integration (VII) proof of concept (POC) test–Executive summary," U.S. Department of Transportation, IntelliDrive(SM), Tech. Rep., Feb. 2009.

[25] A. Vinel, "3GPP LTE versus IEEE 802.11p/WAVE: Which technology is able to support cooperative vehicular safety applications?" *IEEE Wireless Commun. Lett.*, vol. 1, no. 2, pp. 125–128, Feb. 2012.

[26] L. Lei, Y. Zhang, X. Shen, C. Lin, and Z. Zhong, "Performance analysis of device-to-device communications with dynamic interference using stochastic petri nets," *IEEE Trans. Wireless Commun.*, vol. 12, no. 12, pp. 6121–6141, Dec. 2013.

[27] L. Lei, Z. Zhong, C. Lin, and X. Shen, "Operator controlled device-to-device communications in LTE-advanced networks," *IEEE Wireless Commun.*, vol. 19, no. 3, pp. 96–104, Jun. 2012.

[28] "Study on lte device to device proximity services; radio aspects (release 12)," Tech. Rep. 36.843 V12.0.1, Mar. 2014.

[29] Y. Morgan, "Notes on DSRC amp; WAVE standards suite: Its architecture, design, and characteristics," *IEEE Commun. Surveys Tuts.*, vol. 12, no. 4, pp. 504–518, 2010.

[30] G. Karagiannis, O. Altintas, E. Ekici, G. Heijenk, B. Jarupan, K. Lin, and T. Weil, "Vehicular networking: A survey and tutorial on requirements, architectures, challenges, standards and solutions," *IEEE Commun. Surveys Tuts.*, vol. 13, no. 4, pp. 584–616, 2011.

[31] P. Papadimitratos, A. La Fortelle, K. Evenssen, R. Brignolo, and S. Cosenza, "Vehicular communication systems: Enabling technologies, applications, and future outlook on intelligent transportation," *IEEE Commun. Mag.*, vol. 47, no. 11, pp. 84–95, Nov. 2009.

[32] H. Moustafa and Y. Zhang, *Vehicular Networks: Techniques, Standards, and Applications*, 1st ed.   Boston, MA, USA: Auerbach Publications, 2009.

[33] F. Martinez, C.-K. Toh, J.-C. Cano, C. Calafate, and P. Manzoni, "Emergency services in future intelligent transportation systems based on vehicular communication networks," *IEEE Trans. Intell. Transp. Syst. Mag.*, vol. 2, no. 2, pp. 6–20, Oct. 2010.

[34] W.-S. Soh and H. Kim, "QoS provisioning in cellular networks based on mobility prediction techniques," *IEEE Commun. Mag.*, vol. 41, no. 1, pp. 86–92, Jan. 2003.

[35] W. Wanalertlak, B. Lee, C. Yu, M. Kim, S.-M. Park, and W.-T. Kim, "Behavior-based mobility prediction for seamless handoffs in mobile wireless networks," *Wireless Networks*, vol. 17, no. 3, pp. 645–658, Apr. 2011.

[36] G. Remy, S. M. Senouci, F. Jan, and Y. Gourhant, "LTE4V2X-collection, dissemination and multi-hop forwarding," in *Proc. IEEE International Conference on Communications (ICC)*, Ottawa, ON, Jun. 2012, pp. 120–125.

[37] G. Remy, S.-M. Senouci, F. Jan, and Y. Gourhant, "LTE4V2X - impact of high mobility in highway scenarios," in *Proc. Global Information Infrastructure Symposium (GIIS)*, Da Nang, Aug. 2011, pp. 1–7.

Chapter 3
Efficient MAC Mechanisms for Heterogeneous Vehicular Networks

The Medium access control (MAC) layer provides a variety of functions that support the operation of wireless networking. In general, the (MAC) layer manages and maintains communications between either vehicles or vehicle-to-infrastructure by coordinating access to a shared radio channel and utilizing protocols that enhance communications over a wireless medium. This chapter addresses several important issues of MAC layer in vehicular networks, and focuses primarily on channel access control and broadcasting mechanisms. This chapter is organized as follows: Sect. 3.1 presents an overview of channel access protocols in vehicular networks reported in the literatures. Then, the broadcast/multicast protocols are discussed briefly in Sect. 3.2. Section 3.3 presents a location-based channel congestion control mechanism. Finally, insights and discussions are given in Sect. 3.4.

3.1 Channel Access Protocols in Vehicular Networks

A vehicular network has to guarantee safety-related applications first before providing other data services. Safety-related applications, such as the collision avoidance alert, urgent brake alert, and roadside hazards warning, require instant message delivery with a short delay, which should not be interfered with non-safety related applications. Therefore, DSRC with low end-to-end delay is preferred for use among vehicles. One Control channel (CCH) is assigned to transmit critical messages, such as safety warning, while six Service channels (SCHs) are used for various data service applications, such as the point-of-interest notification, and map downloading and updating [1]. Moreover, an efficient multi-channel MAC scheme is essential to guaranteeing the QoS requirements of safety-related applications.

© The Author(s) 2016
K. Zheng et al., *Heterogeneous Vehicular Networks*, SpringerBriefs in Electrical and Computer Engineering, DOI 10.1007/978-3-319-25622-1_3

3.1.1 EDCA in IEEE 802.11p

The EDCA mechanism is employed in DSRC based on IEEE 802.11p. It provides a distinguishing distributed channel access mechanism to guarantee QoS requirements. Four Access category (AC)s are defined in order to support data traffic with different priorities. The distinction between priorities is obtained by setting the values of different parameters such as the Arbitration interframe space (AIFS), minimum contention window (CW_{min}), and maximum contention window (CW_{max}) for different ACs. AIFS is used to indicate the priority of the AC to access the channel. The smaller the AIFS, the higher the chance of transmission. Meanwhile, the Contention Window (CW) size can be selected within the range of $CW_{min} \leq CW \leq CW_{max}$. The shorter the ($CW$) size, the higher the chance to access the channel. Thus, the higher priority AC is assigned with a shorter CW size for a higher probability of channel access, and the data can be possibly transmitted with lower end-to-end latency.

EDCA is a typical (LBT) MAC mechanism. A vehicle has to sense the wireless medium first until a free channel duration of $AIFS$ is completed. If no busy activity is detected in this duration, the transmission starts immediately. Otherwise, a random backoff procedure is to be invoked. Then, the vehicle first randomly generates an integer known as its *Backoff Counter*, which corresponds to the uniform distribution over the interval $[0, CW]$. During the backoff process, carrier sensing (physical&virtual) has to be performed all the time. The *Backoff Counter* decreases by one as long as the medium is free for one time slot duration, i.e., T_{slot}. If the channel becomes busy anytime during the backoff process, the *Backoff Counter* has to be frozen until the channel is clear for an $AIFS$ duration again. Once the *Backoff Counter* is reduced to zero, a transmission can start immediately. Under the ideal condition, collisions do not occur when multiple vehicles choose the same *Backoff Counter*.

The CW is initially set to be CW_{min} and doubles itself each time the previous transmission is unsuccessful. For example, a successful transmission is acknowledged by a correct Acknowledgement (ACK) frame, until it reaches CW_{max}. Once an entire frame is successfully transmitted, the CW is reset to the value of CW_{min}. The two-fold growth mechanism of CW is to ensure the stable performance under overloaded conditions. Another backoff procedure has to be invoked after each transmission so that the transmitted frames from a vehicle are always separated by at least one backoff interval. It can further reduce collisions when the traffic is heavy [2].

3.1.2 Channel Switch in IEEE 1609.4

IEEE 1609.4 defines a management extension to the MAC that allows a system with one or more radios to efficiently switch among them. As shown in Fig. 3.1, channel switching is designed to support data exchanges involving one or more switching devices with a concurrent alternating operation on the CCH and an SCH.

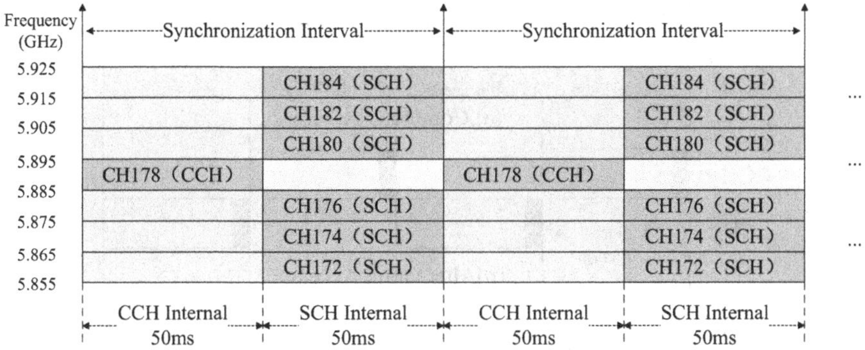

Fig. 3.1 Illustration of the channel switch mechanism in IEEE 1609.4

This allows, for example, a single-PHY device access to high priority data and management traffic during the CCH interval (CCHI), as well as high layer traffic during the SCH interval (SCHI).

One second time is divided into an integer number of synchronization intervals. As safety applications usually require a broadcast rate at about 10 Hz, thus a Sync Interval is set to 100 ms, which is comprised of a CCHI and an SCHI, each lasting for 50 ms in the alternating access mode. Vehicles are synchronized with Universal Coordinated Time (UTC) obtained from sources like the Global Positioning System (GPS). Accounting for the radio switching delay and timer drift among different vehicles, a Guard interval (GI) exists at the beginning of every channel interval. Once the GI reaches the end, usually 4 ms, data transmission may commence. During the GI, no transmission is allowed and the wireless medium is declared as busy to the MAC layer, thus to invoke a random backoff procedure after the GI expires to prevent heavy collisions caused by multiple switching devices attempting to transmit simultaneously at the end of a guard interval. However, when a large number of vehicles have buffered data, heavy collisions are difficult to avoid.

As shown in Fig. 3.2, there are four types of channel accesses, i.e., the continuous access, alternating access, immediate access, and extended access. The continuous access option is rarely used, since it misses the information from either the SCH or CCH. In the normal condition, the basic option, i.e., the alternative access, is adopted. For services which need data intensive transmission, the extended access becomes a better choice, since it improves the transmission rate of the SCH. The immediate access is suited for emergency services due to its immediate switch feature.

The existing channel access methods of IEEE 1609.4 suffer from a few drawbacks, i.e., [3, 4],

1. Channel utilization degrades with the increasing number of vehicles, because more collisions occur due to the CSMA-based MAC scheme adopted;
2. The round-robin concurrent channel switching scheme in IEEE 1609.4 may cause the so-called start-of-interval collision flooding problem. During the CCHI, OBUs that want to send IP data have to wait until an SCHI. Therefore, they tend

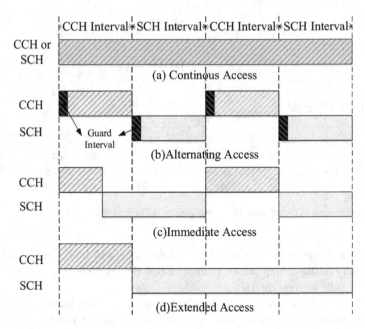

Fig. 3.2 Illustration of channel access in IEEE 1609.4

to transmit messages at the beginning of the SCHI simultaneously, causing many potential collisions;

3. Due to vehicle mobility, the hidden node problem is more severe with a dynamic vehicular topology than in traditional static environments;
4. Frequent Request-to-send (RTS)/Clear-to-send (CTS) exchanges may result in a relatively large overhead; and
5. Due to the limited communications time and the non-association strategy of IEEE 1609.4, an OBU or RSU cannot maintain a status table of its neighbors, making it very difficult and inefficient to implement node-by-node channel access scheduling.

3.1.3 Improved Multi-channel Access Schemes for Vehicular Networks

IEEE 802.11p/1609.4 provides only basic ways of multi-channel operations for vehicular communications. Due to the above problems, much attention has been paid to improve the access performance. In general, there are three types of schemes for reducing the collisions of safety messages and improving data services efficiency. Details are given as follows:

3.1.3.1   Type I: Dedicated Time Slots

A typical scheme to make safety messages more reliable is to dedicate specific resources to safety-related data [5]. For example, in dedicated multi-channel MAC (DMMAC) [5], the channel access time is equally divided into multiple Sync Intervals, each of which consists of a CCHI and an SCHI of the same length. Unlike IEEE 1609.4, the CCHI can be further divided into an Adaptive broadcast frame (ABF) and a Contention-based reservation period (CRP). Moreover, the ABF consists of multiple time slots, which are dynamically reserved by an active vehicle for contention-free delivery of safety messages or other control messages. The slot reservation process is decentralized, similar to R-ALOHA. This provides collision-free and delay-bounded transmissions for safety applications under various traffic conditions.

3.1.3.2   Type II: Coordinator-Based

In dense traffic vehicular networks, a central coordinator managing the local network, similar to the Point Coordination Function (PCF) in IEEE 802.11, can help improve efficiency and performance. In [6], an RSU maintains several tables, which contain information of vehicles associated with the RSU and service requirements of these vehicles, and polls each vehicle one by one in a collision-free period during the CCHI. A Multi-Channel Token Ring Protocol (MCTRP) is designed in [7]. The founder, which initiates a ring, collects and broadcasts emergency messages to the other rings, while a vehicle that wants non-safety data exchange has to wait until the token is passed to it. In the Cluster-Based Multi-Channel Communications Protocol (CBMCCP) [8], the elected CH functions as a coordinator to collect/deliver real-time safety messages within its own cluster, and to forward consolidated safety messages to neighboring CHs. A CH vehicle also controls channel assignment for non-real-time data transmission. In another cluster-based MAC scheme dubbed the Clustering-based Multi-channel MAC Protocol (CMMP) [9], the CH manages the channel status and periodically broadcasts a channel usage list. Before data transmission, each vehicle sends Request Channel Assignment (RCA) in the CCH to the CH in order to obtain available channel resources, which is determined by the CH in consideration of the received requests and channel usage condition.

3.1.3.3   Type III: Contention-Based

This type includes diverse contention-based MAC schemes. In order to reduce collisions and increase channel utilization, the Multi-Channel Beaconing Service (MCBS) is proposed, where the service period is divided into the collision detection and collision avoidance phases [10]. Each vehicle sends beacons with a pre-defined transmission interval specified by the lifetime of a beacon in the collision detection phase. Upon receiving a beacon, each vehicle updates its neighbor table and calculates the probability of a collision. After identifying a potential collision risk, the collision avoidance phase is invoked. In the second phase, it designs a dedicated bi-directional communications process between vehicles. The main objective of the MCBS is to provide benefits for the vehicles in critical situations without interfering with the communications of their neighbors. In [11], the Enhanced multi-channel MAC (EMcMAC) protocol is proposed, where Extended Transmission is designed that allows nodes to extend their services in an SCHI to the CCHI. Safety messages have to be transmitted twice. It greatly improves throughput while still guaranteeing the dissemination of safety message. In [12], a CCH interval is further divided into the safety and WAVE service advertisement (WSA) intervals. Parameters including the CCHIs/SCHIs and the minimum contention window size are dynamically adjusted to ensure high saturation throughput and prioritized transmission of critical safety information. In addition, spacial diversity and cognitive radio techniques can also be used to enhance channel utilization [13, 14].

3.1.4   Summary

Table 3.1 summarizes several MAC schemes for performance improvements in latency and channel utilization for vehicular networks. Various protocols have different requirements on the OBU, e.g., a Single Radio (SR) device or Multi-Radio (MR) device. In a dense vehicular environment, the reservation and coordination process may result in a considerable overhead with the **Type I** and **Type II** schemes. On the other hand, with an increasing number of vehicles, collisions become more severe for the **Type III** scheme.

Table 3.1 Summary of the improved multi-channel MAC protocols

Category	Scheme	Coordinated	Ad-Hoc	SR	MR	Directional	Asyn	Safety	Non-safety
Type I	DMMAC [5]	–	√	√	–	–	–	√	–
Type II	MCTRP [7]	–	√	–	√	–	–	√	√
	CBMCCP [8]	√	–	–	√	–	–	√	√
	CMMP [9]	√	–	–	√	–	–	–	√
Type III	MCBS [10]	–	√	–	√	–	–	√	–
	EMcMAC [11]	–	√	√	–	–	–	√	√
	QVCI [12]	–	√	√	–	–	√	√	–
	VMMAC [13]	–	√	√	–	√	√	–	–
	CROCS [14]	–	√	√	–	–	–	–	√

3.2 Broadcast/Multicast Protocols

Broadcast/multicast services are important in vehicular networks, e.g., road vehicle safety, road navigation support, periodic beacon broadcast, emergency messages dissemination, etc [1]. Therefore, efficient broadcast mechanisms are crucial to minimizing the rate of accidents, enhance traffic efficiency, and improve the travel experience of vehicular users.

Vehicular environments have their unique characteristics in comparison to other wireless networking environments. In a vehicular network, power supply is abundant and mobility trajectory is predictable, which are conducive to implementing broadcast/multicast services [15]. However, there are also several performance-limiting factors such as the large-scale network size, high mobility condition, dynamic network topology, and unreliable connectivity [16].

3.2.1 eMBMS in LTE

LTE can support high-quality multicast and broadcast transmission via the eMBMS functions in the CN and RAN [17]. It is capable of sending data only once to a set of users registered to the offered service, instead of sending it to each node individually. In the LTE system, the transmission of eMBMS data packets is coordinated among a group of tightly synchronized cells, which transmit identical signals on exactly the same time and frequency resources. The signals from these cells are combined over the air, resulting in a strengthened signal. From the terminal perspective, all signals appear to be transmitted from a single large cell. Such a transmission mode is known as the MBMS single frequency network (MBSFN) operation [18].

The eMBMS can be one of possible solutions for the distribution of vehicular services. Traffic management, safety-related and infotainment services can be more efficiently supported by eMBMS in lieu of unicast. In order to guarantee reliable multicast services to each MBMS subscriber, a conservative approach is used for data rate selection. As a consequence, Adaptive modulation and coding (AMC) is provided on a group basis, and the system performance is constrained by the user with the worst channel condition, resulting in increased user dissatisfaction [19]. On the other hand, efficient broadcast and multicast delivery of data may facilitate the development of new services and reinforce the transmission capabilities of current ITS services.

3.2.2 Challenges and Solutions for eMBMS in LTE

Challenges exist when applying eMBMS to message dissemination in HetVNETs. Firstly, since service data are sent by multiple BSs, the propagation delay experienced by the vehicular user may be large. To tackle this issue, an extended cyclic prefix is defined for the eMBMS configuration in LTE, in which only 12 instead of 14 Orthogonal frequency division multiplexing (OFDM) symbols can be transmitted per sub-frame. Moreover, the LTE system is highly susceptible to Inter-carrier interference (ICI) in high mobility scenarios due to large Doppler spreads. Thus, low carrier frequency bands such as 800 MHz with relatively smaller Doppler spreads are recommended for LTE supported vehicular services. On the other hand, due to high mobility, vehicles frequently subscribe to and join in multicast services on a per-user basis. Thus, the signaling overhead increases rapidly. In order to tackle this issue, a dedicated MBMS carrier for downlink only transmission is proposed in [18].

3.2.3 Broadcast Protocols in DSRC

A DSRC system is designed for safety broadcast services. However, when vehicular density is high, the successful message reception probability in IEEE 802.11-based broadcast can be lower than 30 % under saturation conditions [20]. Thus, several challenges remain for existing broadcast protocols to offer reliable and timely services, i.e.,

3.2.3.1 Hidden Node Problem

The RTS/CTS handshake protocol is a mechanism proposed to overcome the hidden node problem. However, broadcast packets have more than one destination, and the RTS/CTS and ACK packets may cause packet storms at the transmitter. Thus, the RTS/CTS handshake is not well suited for broadcast as it is conductive to a more severe hidden node problem than unicast.

3.2.3.2 Fixed Size of Contention Window

The lack of an ACK mechanism causes inability in determining whether a message delivery is successful or not. Thus, regardless of the delivery status, it is impossible to change the CW size for broadcast once the original size is decided. This is the major reason of congestion in broadcast, resulting in a significant reduction in channel utilization [21].

3.2.3.3 Limited Lifetime of Safety Messages

In a vehicular network, beacons carry not only the broadcast information, but also a vehicle's status information. Thus, a beacon contains data from the OBU such as the vehicle speed and location. Due to the highly dynamic network topology, beacon messages are useful only for a limited period of time [22]. If such a message cannot be transmitted before the next beacon is generated, the information it contains becomes invalid. Therefore, this brings in more strict requirements when the broadcast mechanism is adopted in the vehicular network.

3.2.3.4 Broadcast Storm Problem

The broadcast storm is a well-known problem, caused by excessive retransmissions [23]. In vehicular networks, this may happen when vehicles try to send information packets relating to traffic events. To keep the information alive, each vehicle receiving the messages attempts to flood or forward the same packets to all the other vehicles within its coverage range via the CCH/SCH. This leads to large end-to-end latency and low channel utilization. The circumstance becomes worse when the vehicle density is high.

3.2.4 Improved Broadcast Protocols for DSRC

The main objective of broadcast protocols is to provide reliable packet transmission with minimum latency, maximum throughput, and low communication overhead. In accordance with the number of broadcast hops, broadcast protocols can be broadly classified into one-hop broadcast and multi-hop broadcast protocols. They can also be divided into centralized and decentralized broadcast protocols based on the presence or absence of a centralized broadcast node. In this subsection, from the viewpoint of factors that may impact on the forwarding decision, we classify broadcast/multicast protocols into the following categories, i.e., road segmentation-based, link-based, and threshold-based.

3.2.4.1 Road Segmentation-Based Method

Due to rapid network topology variation and node movement in vehicular networks, it is difficult to determine which node is used to forward packets based only on the topology. To tackle this issue, a multi-hop broadcast protocol, termed Urban Multi-hop Broadcast (UMB) is proposed in [24]. This protocol assigns the task of forwarding and acknowledging broadcast packets to only a single vehicle by dividing the road portion inside the transmission range into segments, and choosing the vehicle in the furthest non-empty segment without *a priori* topology information. To guarantee the reliability of multi-hop broadcast, an ACK packet is fed back to the forwarding vehicle, and a mechanism similar to the RTS/CTS handshake is also employed to avoid collisions due to hidden nodes. However, there is usually no Line-of-sight (LoS) link between the vehicles nearby the intersection. To ensure broadcast in this situation, repeaters are installed at intersections to disseminate messages omni-directionally [25]. In this way, packet delivery of high success and efficient channel utilization even with high packet loads are achievable with a fully ad-hoc intersection broadcast mechanism. However, none of these protocols have taken the link status into account when selecting the forwarding vehicle.

3.2.4.2 Link-Based Method

Various factors such as the link status can be considered in choosing a forward delivery node in multi-hop broadcast schemes, e.g., the Link-based Distributed Multi-hop Broadcast (LDMB) scheme [26]. This forward scheme is completely distributed without the need of any handshake. Each vehicle receiving an emergency message first estimates its link status, and then calculates the waiting time before forwarding this message. This scheme takes the link status into consideration, which is concerned mainly with the distance between the sender and receiver, transmission power, transmission rate, and vehicular traffic density. Compared with other multi-hop broadcast protocols, LDMB offers similar performance in reliability while enabling lower latency.

3.2.4.3 Threshold-Based Method

Aside from the link status, other factors such as speed deviation, message priority, and vehicle density may also affect broadcast performance. A threshold can be yielded using various functions to determine whether a node should forward the broadcast message. In [27], a decision threshold function is designed, which simultaneously adapts to the number of neighbors, the node clustering factor, and the Rician fading parameter. In [28], vehicles make their own decisions to forward and acknowledge received packets without a priori topology knowledge. The decision is made according to the distance from the source, the received signal strength indicator (RSSI) of messages, the speed deviation between the sender and forwarder, and the priority of the received messages. However, the performance is still dependent on vehicle density. To solve this problem, a stochastic broadcast scheme is proposed in [29] as an efficient solution to the data dissemination problem. It instructs nodes to rebroadcast messages with a retransmission probability. Unfortunately, choosing a proper retransmission probability is not a simple task. Towards this end, the similarity between stochastic broadcast and the theory of continuum percolation is demonstrated, and the crucial percolation threshold (about 4.5 neighbors on average) in continuum percolation is obtained. Then, the retransmission probability is adjusted so that the apparent density of the network approaches the critical threshold for ensuring greater success with minimum bandwidth.

3.2.5 Summary

Reliable packet transmission with minimum end-to-end delay is the main challenge for designing multi-hop broadcast schemes for HetVNETs. Besides, channel utilization and signal overhead must be taken into account. Meanwhile, a variety of issues, such as the hidden terminal and broadcast storm problems, need urgent attention. In this subsection, we investigated some existing multi-hop broadcast protocols for vehicular networks that are divided into three groups, i.e., the road segmentation-based, link-based, and threshold-based methods, as shown in Table 3.2. Most of the above protocols solve only part of these issues with some limitations. For example, topology information is imperative for the road segmentation-based method. However, accurate topology information is difficult to obtain in a vehicular environment. A fast moving vehicle has difficulties in accurately estimating the

Table 3.2 Summary of broadcast protocols in vehicular networks

	Category	Broadcast protocol	Hand shake	Service	Typical deployment	Feature
LTE	eMBMS [18]	eMBMS [18]	×	Safety and traffic efficiency services	Highway, urban	Centralized architecture, mobility management, high capacity, and large coverage
DSRC	Road-segmentation-based method	UMB [24]	√	Not mentioned	Urban, nonline-of-sight intersection	Need repeaters installed at the intersection
		AMB [25]	√	Not mentioned	Urban, nonline-of-sight intersection	Vehicle as repeater at the intersection
	Link-based method	LDMB [26]	×	Emergency/safety services	Not mentioned	Link status is considered
	Threshold-based method	DADCQ [27]	×	Not mentioned	Highway, urban	Adaptive to node density and distribution, channel quality
		HMB [28]	√	Safety services	Highway	Passive forwarder selection and acknowledgment
		Stochastic broadcast [29]	×	Not mentioned	Not mentioned	Adjust the retransmit probability according to node density

link quality, which impacts on the performance of the link-based method. With regards to the threshold-based method, vehicle density is a performance bottleneck. Furthermore, the performance and problems when eMBMS is applied in vehicle networks were also discussed. The signaling overhead due to frequent joining in and leaving eMBMS services, and the large propagation delay caused by multiple-eNBs coordinated broadcast have to be dealt with. How to take advantage of both DSRC and LTE technologies while tackling these challenging issues remains a hard "nut" to crack.

3.3   Location-Based Channel Congestion Control Mechanism

Since there already exist lots of reported studies on message dissemination in the CCH [4, 30], we pay more attention to the SCH performances in this section. Due to numerous non-safety applications in vehicular networks, frequent exchanges of large packets are prevailing, which may lead to a high degree of network congestion. Thus, the efficiency of channel access in the SCH is very crucial in V2I links. The V2I communications based on IEEE 802.11p are usually under *high-speed high-density* environments, e.g., hundred of vehicles moving at the speed up to 120 km/h and communicating with the RSUs. Therefore, the existing channel access schemes of IEEE 802.11p/1609.4 V2I links suffer from a few drawbacks [31, 32].

On the other hand, vehicular environments have their own characteristics that can be exploited to improve the network performance. For example, the trajectory of vehicles is usually fixed, and their location and speed information is accessible through devices such as GPS. Therefore, we propose a location-based channel congestion control mechanism in the MAC layer to enhance the access performance in the SCH while keeping the backwards compatibility [33]. Vehicles are firstly divided into groups with separate channel access time windows based on their instantaneous location information, which greatly reduces the start-of-interval collisions. The grouping method is based on the road section allocation determined by the RSUs, thus no node-by-node time slot scheduling is required. Moreover, the RSUs can also dynamically adjust the road section size to adapt to different traffic load. Then, the vehicles with less time in the network have a higher priority, and thus the overall system fairness is improved. The hidden node problem is also properly solved because remotely apart vehicles are within different time windows.

3.3.1   Location Segmentation

The road within the coverage of an RSU is divided into N geographical sections as illustrated in Fig. 3.3, denoted as $\mathbb{S} = \{S_1, S_2, \cdots, S_n\}$. The length of the sections are $\mathbb{L} = \{L_1, L_2, \cdots, L_n\}$. A reference direction is chosen as the road driving direction, which is indicated by the Reference Direction Arrow in Fig. 3.3. At any time, a vehicle is in one of the sections.

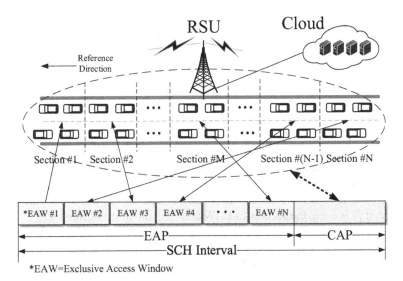

Fig. 3.3 Illustration of the location-based channel congestion control mechanism in V2I communications

3.3.2 Improved SCH Intervals Structure

The SCHI is divided into two periods, i.e., an Exclusive access period (EAP) and a Contention access period (CAP) with the time durations of T_E and T_C, respectively. In EAP, channel access complies with a Time-division multiple-access (TDMA)/CSMA rule to spread out start-of-channel-access time, which will be presented in detail later. In CAP, the vehicles whose transmissions failed in EAP can compete the channel by using the EDCA defined in IEEE 802.11p, which can either improve the fairness or provide urgent message dissemination.

The EAP is further partitioned into multiple Exclusive access windows (EAWs) with the time duration denoted as $\mathbb{W} = \{W_1, W_2, \cdots, W_m\}$. A vehicle is bound to a specific EAW, in which it contends for data transmission, according to which road section it currently belongs to.

3.3.3 Section-to-EAW Mapping

A mapping function from the road section, $\mathbb{S}$ to the EAW $\mathbb{W}$, i.e., $f : \mathbb{S} \to \mathbb{W}$, is designed to spread the attempting access time of vehicles aiming at reducing the collisions. Previous research work has not taken the vehicles' sojourn time in the network into account. In V2I vehicular communications, as vehicles drive through the RSU coverage area, the remaining communications time becomes less, and thus data transmission for the vehicles in the front of the fleet is more stringent.

Table 3.3 Example table of section-to-EAWs mapping

EAW Index($W(i)$)	1	2	3	4	$\cdots$
Section Index(i)	1	N	2	$N-1$	$\cdots$

Thus, the essential idea behind this proposed MAC is to allocate the time window according to the remaining time for vehicles to communicate in the network. Generally, the vehicles in the leading position have less communications time left, so they need to have a higher priority. We propose to assign sections further from the RSU to the earlier EAWs in the SCHI. As there may be several parallel lanes with opposite driving directions, we alternatively select the sections on the opposite sides and map them to the sequential EAWs.

Assume that both the road and EAP are divided into N equal partitions. For instance, if N is even, the section-to-EAW mapping function is given by

$$W(i) = \begin{cases} 2i - 1, & i < \frac{N}{2} + 1 \\ 2(N + 1 - i), & i \geq \frac{N}{2} + 1 \end{cases} \tag{3.1}$$

where i is the index of Section $\mathbb{S}$, and $W(i)$ is the corresponding index of the EAW $\mathbb{W}$. Then, the mapping table is shown in Table 3.3. The RSU broadcasts the information of the road sections, EAW partitions, and the mapping table in the CCHI, by using a Vendor Specific Action frame with Exclusive channel access (ECA), to all the nearby vehicles. Meanwhile, vehicles check and find their own EAW according to the location and then employ proper channel access.

3.3.4 Modified CSMA/CA Scheme

When a massive number of vehicles exist in the fixed length SCHI, it is inevitable that several vehicles are assigned to the same EAW. Here a modified CSMA/CA scheme is applied for vehicles in the same EAW to contend for channel.

The vehicles within the same EAW are designated as the owners of this EAW, while others are regarded as non-owners. As vehicles drive through different sections, they are allocated into different EAWs. The basic rules of the channel access mechanism in the EAP, dubbed ECA, are described as follows:

- *Rule 1*: A vehicle can only transmit in the specific EAW that it owns. In other words, an EAW is exclusively used by its owners and restricted from access by non-owners;
- *Rule 2*: The vehicles within an EAW use EDCA to contend for channel access. A regular random backoff procedure needs to be performed when an IDLE channel of the *AIFS* length is detected for the first time to avoid concurrent start-of-EAW collisions; and
- *Rule 3*: The vehicles departing from the RSU are prioritized over those approaching it. A vehicle can judge whether it is leaving or approaching the

RSU by its moving direction and the relative position variation with the RSU. The vehicles in section(s) close to the RSU are of the same priority. Those sections are explicitly indicated in the ECA broadcast frame sent by the RSU in the previous CCHI. The vehicles that are departing from the RSU have initial contention window set to CW_{min} by default, while approaching vehicles set to $2CW_{min} - 1$. In this way, vehicles with less communications time have a higher priority than those have more time.

Since an EAP is slotted into multiple EAWs, the transmission may be across the slot boundary, which can be used to increase channel utilization. When the basic ECA rules are designed to spread out the potential collisions in dense networks, they suffer from the loss of channel utilization in the case of low traffic load in the network. Thus, an optional Prioritized exclusive channel access (PECA) scheme is designed to tackle this problem, enabling non-owners to contend for channels. The vehicle as the owner contend for channel access with the basic ECA scheme, while that as the non-owner can perform the following steps:

- *Step 1*: It first listens to the channel from the start of an EAW until it detects a free channel during the *AIFS*;
- *Step 2*: When the first *AIFS* free channel time is reached, it listens to the channel for another CW_{min} duration. If any transmission is detected during the CW_{min} period, it has to abort the transmission attempt in this EAW; and
- *Step 3*: When a clear channel of length CW_{min} is detected as stated in *Step 2*, it may invoke a common random backoff procedure for data transmission with the initial *CW* of $2CW_{min} - 1$.

Since the PECA scheme is designed to improve channel utilization only in the event of low traffic, it can be used unless the RSU explicitly enables it in the ECA frame. The RSU may determine whether or not to enable it according to the network traffic load monitored.

3.3.5 Frame Structure for ECA

A sample of the Vendor Specific Information Element [34], termed the ECA element, is given in Fig. 3.4 to illustrate how the RSU broadcasts necessary information in the proposed scheme. The EAP control field contains the information about the EAP length and PECA capability. The EAW information field describes what information comprises EAWs, e.g., the number, duration, and so on. The last four fields formulate road sections information and the section-to-EAW mapping table content. If the EAWs or road sections are unequally divided, the optional fields in the dashed box may be needed. The RSU broadcasts the frame containing such information element in the CCHI. Then, the vehicles decode it and perform channel access in the SCHI as demanded.

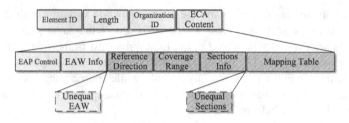

Fig. 3.4 Sample of ECA information element

The proposed ECA method is scalable as the section-to-EAW mapping is completely carried out at the RSU side, which requires no node-by-node scheduling. The hidden node problem is greatly alleviated since the vehicles in the same EAW are geographically close to each other. Furthermore, as the RSU can monitor the network traffic load measured by the collision ratio or data queue length, it can dynamically adjust the number of road sections to achieve a stable performance.

3.3.6   Performance Analysis

Assume that vehicles in each lane move towards the same direction with the same mean velocity v, passing the RSU along the two-lane road. The RSU has a coverage area with a radius of R_{RSU}. For simplicity but without loss of generality, vehicles are uniformly distributed in one lane. The road is divided into several sections, i.e., $\mathbb{S} = \{S_1, S_2, \cdots, S_N\}$. When a vehicle leaves the RSU coverage from the end of section S_1, it is assumed to re-enter the road from the start section S_N, or vice versa. The main purpose of the proposed scheme is to deal with the start-of-interval collision flooding problem. Therefore, our simulations are carried out in a saturated condition, where all the vehicles always have packets to send with the same packet length $p_L = 1000$ bytes of the same AC. The perfect channel condition is assumed in our simulations so that no transmission errors occur, similar to those in [35, 36]. In our evaluation, the section size can be set to be 500, 200, 100, and 50 m, which results in the different numbers of sections and EAWs, i.e., 2, 5, 10, and 20, respectively. The detailed simulation parameters are listed in Table 3.4.

Figure 3.5 plots the collision ratios of two channel access schemes, i.e., the one defined by the default IEEE 802.11p MAC and the proposed one. It turns out that the proposed scheme can greatly reduce the occurrence of collisions, especially when the network becomes congested. For example, the collision ratio of the proposed scheme with 20 EAWs is only less than 50 % of that of the default channel access scheme. Moreover, with the increase in the number of road sections, i.e., shorter but more numbers of EAWs in the EAP, collisions are inclined to drop. This is because less vehicles are assigned to the same time windows contenting for channel access with less collisions, when the EAW number increases. Although a larger number of road sections lead to a less number of collisions, it cannot always be increased due

Table 3.4 Simulation parameters

Parameter	Value
$aSlotTime$	$10\,\mu s$
$SIFS/AIFS$	$30\,\mu s/50\,\mu s$
CW_{min}/CW_{max}	15/1023
R_{RSU}	500 m
Packet load (p_L)	1,000 Bytes
Data rate	3 Mbps
Vehicle velocity	80 km/h

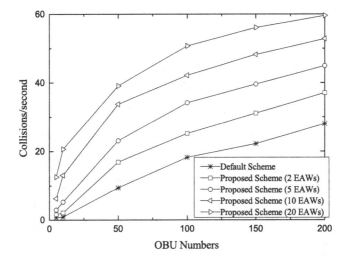

Fig. 3.5 Performance comparison on the collision ratio

to limitations such as the vehicle length, car space, GPS precision, etc. Besides, as the SCHI is usually fixed to 50 ms for each packet (1,000 bytes) and about 3 ms is needed to send at the lowest rate, 20 EAWs are fairly large so that each EAW barely holds a duration of per packet transmission time. Another important limiting factor is due to the corresponding overhead to transmit in the ECA broadcast frame during the CCHI.

The average packet delivery delay is also given in Fig. 3.6. The delay increases drastically when the number of vehicles is too large. Compared with the default channel access scheme in IEEE 802.11p, the proposed scheme can reduce the average delay by 21 % due to less collisions.

Figure 3.7 compares the system throughputs of the two channel access schemes. The proposed scheme is able to improve the system throughput. Vehicles do not perform channel access simultaneously but in different time windows by using the proposed scheme. As collisions in each small window is much less likely than that over the entire interval, there will be more packets successfully transmitted, attributing to the better throughput performance.

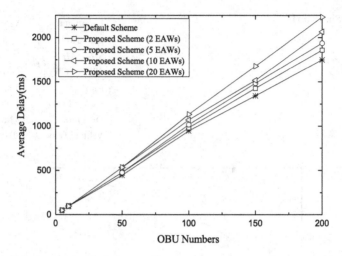

Fig. 3.6 Performance comparison on the average packet delivery delay

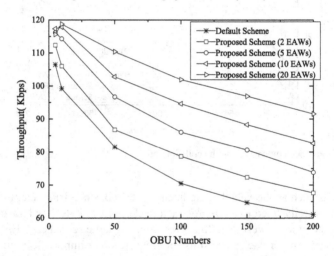

Fig. 3.7 Performance comparison on the system throughput

3.3.7   Summary

To tackle the collision flooding problem caused by the CSMA/CA-based periodic
channel switch mechanism in the dense IEEE 802.11p/1609.4 V2I network, a
location-based TDMA/CSMA MAC scheme was proposed in this section. Vehicles
are divided into different channel access windows according to their geographical
locations, which help spread out potential packet collisions over the entire SCHI.
Forward driving vehicles, which have less communications time, are given a higher
priority for channel access to ensure time stringent transmission. The proposed

scheme is scalable and may adapt to the current network traffic load. The hidden node problem is also resolved since remote nodes have different access time windows. Besides, the proposed scheme can be easily extended beyond the existing IEEE 802.11p/1609.4 protocols. Simulation results demonstrated that our scheme is able to outperform the default channel access scheme defined in IEEE 802.11p MAC, both in delay reduction and system throughput improvement in dense scenarios.

3.4 Insights and Discussions

Due to the co-existence of multiple wireless communications techniques in the HetVNET, the integration of different MAC protocols and functions associated different techniques is a challenging issue. As such, a new HLL is introduced on the top of the MAC layers in different radio access systems to make these layers work collaboratively. This new layer can jointly manage the radio resources across a variety of systems, and balance the traffic among various networks. The development of Network Function Virtualization (NFV) techniques provides the feasibility to implement the HLL functions.

In an HetVNET, it is expected to have contention-based channel access mechanisms as well as contention-free ones, which may depend on specific services, e.g., contention-free mechanisms for media download. Contention-based mechanisms require no coordination, and thus are robust to network topology change and of low overhead in the event of low or medium traffic loads. The contention-based MAC is more suitable for local area services, such as abnormal condition warnings and emergency vehicle warnings. On the other hand, the contention-free MAC mechanisms possess the advantages such as the guaranteed QoS, high efficiency in the presence of heavy traffic loads. They are more suitable for wide-area coverage to all vehicles in a large-scale network. Therefore, an elegant trade-off between the contention-free and contention-based channel access mechanisms needs to be investigated in HetVNETs.

Another challenge in designing MAC in the HetVNET stems from the contradiction between the strict latency requirement and the extra delay due to the management across different systems. Such a problem becomes serious especially in high mobility environments. To support safety-related services, the network efficiency needs to be traded for reliability when studying new MAC schemes.

Since various emerging vehicular services have their own requirements, how to guarantee appropriate levels of QoS for these services in the HetVNET becomes a challenging issue. Each wireless communications technique in the HetVNET has its specific features to support certain services. Then, all the radio resources in the HetVNET can be jointly exploited through joint radio resource management in the MAC layer.

Finally, it is highly suggested that the characteristics of the specific application scenarios in the HetVNET be taken into consideration, e.g., highway, platooning, and so on. Such scenarios have different traffic and communication features, which may raise new challenges to the design of MAC schemes.

References

[1] K. Zheng, Q. Zheng, P. Chatzimisios, W. Xiang, and Y. Zhou, "Heterogeneous vehicular networking: A survey on architecture, challenges and solutions," *IEEE Commun. Surveys Tuts.*, vol. PP, no. 99, pp. 1–1, 2015, DOI: 10.1109/COMST.2015.2440103.

[2] "IEEE Standard for Information Technology–Telecommunications and Information Exchange Between Systems Local and Metropolitan Area Networks–Specific Requirements Part 11: Wireless LAN Medium Access Control (MAC) and Physical Layer (PHY) Specifications," *IEEE Std 802.11-2012 (Revision of IEEE Std 802.11-2007)*, pp. 1–2793, Mar. 2012.

[3] "IEEE Standard for Wireless Access in Vehicular Environments (WAVE) Multi-channel Operation," *IEEE Std 1609.4-2010 (Revision of IEEE Std 1609.4-2006)*, pp. 1–89, 2011.

[4] Q. Wang, S. Leng, H. Fu, and Y. Zhang, "An IEEE 802.11p-based multichannel MAC scheme with channel coordination for vehicular ad hoc networks," *IEEE Trans. Intell. Transp. Syst.*, vol. 13, no. 2, pp. 449–458, Jun. 2012.

[5] N. Lu, Y. Ji, F. Liu, and X. Wang, "A dedicated multi-channel MAC protocol design for VANET with adaptive broadcasting," in *Proc. IEEE Wireless Communications and Networking Conference (WCNC)*, Sydney, Apr. 2010, pp. 1–6.

[6] S.-T. Cheng, G.-J. Horng, and C.-L. Chou, "Using cellular automata to form car society in vehicular ad hoc networks," *IEEE Trans. Intell. Transp. Syst.*, vol. 12, no. 4, pp. 1374–1384, Jun. 2011.

[7] Y. Bi, K.-H. Liu, X. Shen, and H. Zhao, "A multi-channel token ring protocol for inter-vehicle communications," in *Proc. IEEE Global Telecommunications Conference (GTC)*, New Orleans, Nov. 2008, pp. 1–5.

[8] X. Zhang, H. Su, and H.-H. Chen, "Cluster-based multi-channel communications protocols in vehicle ad hoc networks," *IEEE Wireless Commun.*, vol. 13, no. 5, pp. 44–51, Oct. 2006.

[9] T. Kim, S. Jung, and S. Lee, "CMMP: clustering-based multi-channel MAC protocol in VANET," in *Proc. Second International Conference on Computer and Electrical Engineering (ICCEE)*, vol. 1, Dubai, UAE, Dec. 2009, pp. 380–383.

[10] R. Lasowski and M. Strassberger, "A multi channel beaconing service for collision avoidance in vehicular ad-hoc networks," in *Proc. IEEE Vehicular Technology Conference (VTC)*, San Francisco, Sep. 2011, pp. 1–5.

[11] Q. Wang, S. Leng, H. Fu, Y. Zhang, and H. Weerasinghe, "An enhanced multi-channel MAC for the IEEE 1609.4 based vehicular ad hoc networks," in *Proc. IEEE INFOCOM Workshops*, San Diego, Mar. 2010, pp. 1–2.

[12] Q. Wang, S. Leng, Y. Zhang, and H. Fu, "A QoS supported multi-channel MAC for vehicular ad hoc networks," in *Proc. IEEE Vehicular Technology Conference (VTC)*, Budapest, Hungary, May 2011, pp. 1–5.

[13] X. Xie, B. Huang, S. Yang, and T. Lv, "Adaptive multi-channel MAC protocol for dense VANET with directional antennas," in *Proc. IEEE Consumer Communications and Networking Conference (CCNC)*, Las Vegas, NV, Jan. 2009, pp. 1–5.

[14] J.-H. Chu, K.-T. Feng, J.-S. Lin, and C.-H. Hsu, "Cognitive radio-enabled optimal channel-hopping sequence for multi-channel vehicular communications," in *Proc. IEEE Vehicular Technology Conference (VTC)*, San Francisco, Sep. 2011, pp. 1–5.

[15] H. Song and H. S. Lee, "A survey on how to solve a decentralized congestion control problem for periodic beacon broadcast in vehicular safety communications," in *Proc. International Conference on Advanced Communication Technology (ICACT)*, Pyeongchang, Jan. 2013, pp. 649–654.

[16] H. Moustafa and Y. Zhang, *Vehicular Networks: Techniques, Standards, and Applications*, 1st ed. Boston, MA, USA: Auerbach Publications, 2009.

[17] G. Araniti, C. Campolo, M. Condoluci, A. Iera, and A. Molinaro,, "LTE for vehicular networking: A survey," *IEEE Commun. Mag.*, vol. 51, no. 5, pp. 148–157, May 2013.

[18] "Intelligent transport systems (ITS); framework for public mobile networks in cooperative ITS (C-ITS)," European Telecommunications Standards Institute (ETSI), Tech. Rep. 102 962 V1.1.1, Feb. 2012.

[19] G. Araniti, V. Scordamaglia, M. Condoluci, A. Molinaro, and A. Iera, "Efficient frequency domain packet scheduler for point-to-multipoint transmissions in LTE networks," in *Proc. IEEE International Conference on Communications (ICC)*, Ottawa, Jun. 2012, pp. 4405–4409.

[20] M. T.-M. D. Jiang and H. Hartenstein., "Broadcast reception rates and effects of priority access in 802.1 i-based vehicular ad-hoc networks," in *Proc. ACM VANET04*, Philadelphia, Oct. 2004.

[21] J. Alapati, B. Pandya, S. Merchant, and U. Desai, "Back-off and retransmission strategies for throughput enhancement of broadcast transmissions in 802.11p," in *Proc. IEEE Intelligent Vehicles Symposium (IV)*, San Diego, Jun. 2010, pp. 700–705.

[22] R. Stanica, E. Chaput, and A. Beylot, "Properties of the MAC layer in safety vehicular ad hoc networks," *IEEE Commun. Mag.*, vol. 50, no. 5, pp. 192–200, May 2012.

[23] M. Booysen, S. Zeadally, and G.-J. van Rooyen, "Survey of media access control protocols for vehicular ad hoc networks," *IET Commun.*, vol. 5, no. 11, pp. 1619–1631, Jul. 2011.

[24] G. Korkmaz, E. Ekici, F. Özgüner, and Ü. Özgüner, "Urban multi-hop broadcast protocol for inter-vehicle communication systems," in *Proc. ACM International Workshop on Vehicular Ad-hoc Networks*, New York, 2004, pp. 76–85.

[25] G. Korkmaz, E. Ekici, and F. Ozguner, "Black-burst-based multihop broadcast protocols for vehicular networks," *IEEE Trans. Veh. Technol.*, vol. 56, no. 5, pp. 3159–3167, Sep. 2007.

[26] Q. Yang and L. Shen, "A multi-hop broadcast scheme for propagation of emergency messages in VANET," in *Proc. IEEE International Conference on Communication Technology (ICCT)*, Nanjing, Nov. 2010, pp. 1072–1075.

[27] M. Slavik and I. Mahgoub, "Spatial distribution and channel quality adaptive protocol for multihop wireless broadcast routing in VANET," *IEEE Trans. Mobile Comput.*, vol. 12, no. 4, pp. 722–734, Feb. 2013.

[28] M. Barradi, A. Hafid, and S. Aljahdali, "Highway multihop broadcast protocols for vehicular networks," in *Proc. IEEE International Conference on Communications (ICC)*, Ottawa, Jun. 2012, pp. 5296–5300.

[29] M. Slavik and I. Mahgoub, "Stochastic broadcast for VANET," in *Proc. IEEE Consumer Communications and Networking Conference (CCNC)*, Las Vegas, Jan. 2010, pp. 1–5.

[30] K. J. Song, C. H. Lee, M. S. Woo, and S. G. Min, "Distributed periodic access scheme (DPAS) for the periodic safety messages in the IEEE 802.11p WAVE," in *Proc. International Conference on Communications and Mobile Computing (CMC)*, Qingdao, Apr. 2011, pp. 465–468.

[31] "IEEE Standard for Wireless Access in Vehicular Environments (WAVE)–Multi-channel Operation," *IEEE Std 1609.4-2010 (Revision of IEEE Std 1609.4-2006)*, pp. 1–89, Feb. 2011.

[32] J. Zhu and S. Roy, "MAC for dedicated short range communications in intelligent transport system," *IEEE Commun. Mag.*, vol. 41, no. 12, pp. 60–67, Dec. 2003.

[33] Z. Zeng, H. Zhao, X. Xin, and K. Zheng, "A novel location-based channel congestion control scheme in V2I networks," in *Proc. International ICST Conference on Communications and Networking in China (CHINACOM)*, Guilin, Aug 2013, pp. 118–123.

[34] "IEEE Standard for Information Technology– Local and Metropolitan Area Networks–
     Specific Requirements– Part 11: Wireless LAN Medium Access Control (MAC) and Physical
     Layer (PHY) Specifications Amendment 6: Wireless Access in Vehicular Environments,"
     *IEEE Std 802.11p-2010 (Amendment to IEEE Std 802.11-2007 as amended by IEEE Std
     802.11k-2008, IEEE Std 802.11r-2008, IEEE Std 802.11y-2008, IEEE Std 802.11n-2009, and
     IEEE Std 802.11w-2009)*, pp. 1–51, Jul. 2010.
[35] G. Bianchi, "Performance analysis of the IEEE 802.11 distributed coordination function,"
     IEEE J. Sel. Areas Commun., vol. 18, no. 3, pp. 535–547, Mar. 2000.
[36] T. Luan, X. Ling, and X. Shen, "MAC in motion: impact of mobility on the MAC of drive-thru
     internet," *IEEE Trans. Mobile Comput.*, vol. 11, no. 2, pp. 305–319, Feb. 2012.

Chapter 4
Resource Allocation in Heterogeneous Vehicular Networks

Vehicular networks are facing an overwhelming growth in data traffic demands recently. However, radio resources in wireless networks infrastructures have not been fully exploited, resulting in low quality of services for vehicle users. As a result, efficient radio resource allocation schemes for HetVNETs are in urgent demand. In this chapter, we first present a brief overview on radio resource allocation in vehicular networks. Then, Sect. 4.2 presents a new content-based resource scheduling mechanism. A Bipartite graph (BG)-based cooperative scheduling scheme is also studied in Sect. 4.3, followed by concluding remarks of the chapter in Sect. 4.4.

4.1 Related Work

There exists a plethora of existing studies on radio resource allocation reported in the literature. Most studies aim at traditional heterogeneous networks, such as LTE-based networks. In [1], the state-of-the-art on heterogeneous networks is summarized, and a high-level overview of radio resource management for LTE heterogeneous networks is given. Also, a flow-based framework is formulated for the joint optimization of resource allocation and user association in heterogeneous networks [2]. To deal with the unprecedented challenges imposed by 5G communications ecosystems, emerging heterogeneous network architectures are able to accommodate the integration between multiple radio access techniques [3]. A comprehensive mathematical method for real-time performance optimization is also provided for heterogeneous cloud radio access networks in [3]. It strikes a flexible balance between throughput and fairness. All these works provide valuable insights into radio resource allocation in HetVNETs.

© The Author(s) 2016
K. Zheng et al., *Heterogeneous Vehicular Networks*, SpringerBriefs in Electrical and Computer Engineering, DOI 10.1007/978-3-319-25622-1_4

Other approaches are based on the content that vehicular users request, i.e., the content-based radio resource allocation schemes. They can well support efforts to user classes, properties, and relationships concerning context information. In [4], a novel approach is proposed to cluster the interests of car drivers, increasing the lifetime of the interest group and the throughput of the V2V environment. The essence of this approach is based on the ontology of the user interest. In [5], the applications are firstly classified into different types according to their content. Then, a novel Inter-vehicular communication (IVC) architecture is proposed to adapt its functionalities to serving various applications, ranging from safety messages to entertainment information through granting different priorities to each type.

Meanwhile, cooperative communications are one of the most promising techniques for improving the system performance in wireless communication networks. It is able to reduce the transmission distance and to increase the number of users under favorable channel conditions, leading to better channel quality and higher throughput. This is highly desired in the HetVNET because each of the current wireless technologies has its respective advantages and limitations. Due to the differentiating characteristics and QoS requirements of ITS services, it requires specific cooperative schemes to well support these services. The basic principle of cooperative communications in the HetVNET is that a vehicular user can be assisted by neighboring vehicles or the base station to improve its transmissions. Cooperative mechanisms can enhance the overall performance compared to their direct transmission counterparts. As such, a great deal of research has been undertaken to study cooperative mechanisms in vehicular networks. For example, inter-vehicle communications can be assisted by an RSU with Amplify-and-forward (AF) relaying [6]. Cross-layer approaches for cooperative diversity networks are also studied in [7, 8]. In this chapter, we give a short survey on cooperative communications in vehicular networks from the viewpoints of both decentralized and centralized control.

- **Decentralized Cooperative Communications**: In decentralized cooperative schemes, there lacks a centralized coordinator (e.g., BS or RSU) to schedule the communications links, select the relaying node, and manage the radio resources. Usually a self-organized mechanism is employed, which takes node properties into consideration so as to enable network self-organization. In [9], a novel cooperation protocol is proposed, where the RSUs are enabled to maximize the revenues. Through the cooperation, the RSUs can diversify the service data that they transmit to their service vehicles, depending on the content-sharing likelihood of the underlying V2V network connecting the vehicles. Also, another cooperative communications scheme on link scheduling is proposed in cognitive vehicular networks [10]. A Three-dimensional (3D) graph is used to characterize the conflict of cooperative links. Then, a near-optimally solution is derived through linear programming, which is effective in improving the end-to-end throughput. Moreover, a novel protocol, dubbed the Vehicular cooperative media access control (VC-MAC), is proposed in [11] in an effort way to facilitate the cooperative communications in vehicular networks. Spatial diversity and user diversity are exploited to improve the system throughput and reduce the vulnerability of wireless transmission by cooperative relaying Multiple input multiple output (MIMO) schemes, which are investigated in [12]. The performances of the throughput and energy consumption of cooperative relaying MIMO are also compared with those of traditional techniques.

• *Centralized Cooperative Communications*: Centralized cooperative communications prove efficient in distributing and collecting data to support the ITS services. In a centralized cooperative mechanism, a central node (e.g., BS or RSU) has global control of the vehicles in its coverage area. Based on the global information, a cooperative scheme, e.g., link scheduling and resource allocation, can be derived. This allows for efficient utilization of radio resource and avoidance of interference from other links. A novel centralized framework, termed LTE4V2X, uses both IEEE 802.11p and LTE to collect data periodically from vehicles, and send them back to the central server [13]. In the framework, IEEE 802.11p is used to forward the data from vehicles to BSs cooperatively. On the other hand, V2V communications are an emerging technology that can enhance the connectivity in vehicular networks, and provide an efficient way for cooperative communications among connected vehicles. Thus, an enhanced connectivity scheme using V2V communications is proposed in [14], where IEEE 802.11p is used for short range communications between vehicles, and the LTE system is used for communications between the vehicles and BSs over large range cellular links. In the event that V2I and V2V communications work in the same frequency bands, the co-channel interference between V2I and V2V communications can cause severe system performance degradation. To tackle this challenging issue, beamforming can be used in conjunction with the corresponding resource allocation scheme [15]. In this way, the co-channel interference can be mitigated, resulting in improved overall system capacity. Meanwhile, in order to effectively support ITS services, several new physical layer techniques are introduced for cooperative communications in vehicular networks [7, 8, 16].

4.2 Content-Based Resource Scheduling Mechanism

The predictable delay in contention-based mechanisms such as CSMA in IEEE 802.11p poses great challenges to guaranteed QoS, particularly to providing reliable broadcast support in a congested condition. Thus, a TDMA-based slotted MAC protocol may be employed to share the radio resources in vehicle communication networks.

According to the analysis of user convergence behaviors, multiple vehicles in the same highway region within a short time interval share a common interest in certain applications, such as map download, parking payments, automatic tolling services, collision warning ahead, etc. However, in the DSRC system, safety messages and some traffic management messages are broadcast in the CCH, while most

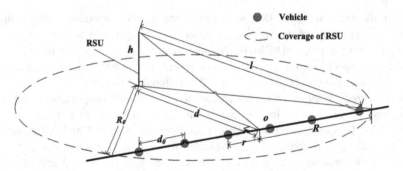

Fig. 4.1 Illustration of a highway scenario

infotainment services are unicast in the SCHs. Such unicast transmission on the SCHs inevitably causes some waste of radio resources, since duplicated contents may be transmitted independently to different requesting vehicles. To solve this problem, the traffic is suggested to be firstly classified into two types according to content similarities, i.e., either shared traffic with the same content or dedicated traffic with different content [17]. Then, these two types of traffic are either broadcast or unicast. The performance of this content-based resource scheduling mechanism can be analyzed by virtue of queueing theory.

4.2.1 System Model

A highway scenario with one lane for one direction is illustrated in Fig. 4.1, where downlink communications are under consideration. The RSU has the transmission range S with a radius of R_0, where N vehicles are served. Without loss of generality, the vehicles are assumed to be distributed uniformly with an equal headway d_0 between any adjacent pair with the same and constant velocity.

The average received Signal-to-noise radio (SNR) of a vehicle varies due to the variation of its distance away from the RSU. So, one needs to firstly compute the distance distribution of vehicles from the RSU in the coverage area, which can then be used to analyze the location/time-averaged transmission rate of vehicles. A list of important notations in our model is summarized in Table 4.1. It is noted that one vehicle is regarded as one user in our analysis. Moreover, subscripts μ and b denote unicast and broadcast transmission, respectively.

4.2.1.1 Unicast Transmission

A coordinate axis with the origin point O is established, which is the intersection point of the mid-perpendicular from the RSU to the road and the middle line of the road. l and r denote the actual distances from the vehicle to the RSU and to point O,

Table 4.1 Notation and assumptions

Expression	Meaning
h	The height of RSU
d	The vertical dimension between the RSU and the road
R	Equivalent coverage radius of the RSU on road
R_0	Coverage radius of the RSU
d_0	Distance between two adjacent vehicles
N	Number of vehicles in the coverage of the RSU
O	The intersection point of d and the middle line of the road

respectively. Since all the vehicles are uniformly distributed on the road and each moves at the same constant speed, the distance r follows a uniform distribution, i.e.,

$$f(r) = \frac{1}{2R}, \quad -R \leq r \leq R. \tag{4.1}$$

Through geometric calculation, the Probability density function (PDF) of the distance between the vehicle and the RSU can be shown as

$$f(l) = \frac{l}{2R}\sqrt{l^2 - d^2 - h^2}, \tag{4.2}$$

where $l \in \left[\sqrt{d^2 + h^2}, \sqrt{d^2 + h^2 + R^2} \right]$.

4.2.1.2 Broadcast Transmission

The transmission rates for unicast and broadcast transmissions are different. The unicast transmission rate depends on the specific channel condition of the only scheduled vehicle, while the QoS requirements of all the other vehicles in broadcast transmission have to be jointly guaranteed. So, the vehicle with the worst channel condition needs to be analyzed when computing the broadcast transmission rate. Moreover, since the average SNR depends mainly on the large-scale fading prorogation, the farthest vehicle from the RSU has the worst channel condition in most cases. However, it is very difficult to identify the farthest vehicle from the RSU instantaneously, when all the vehicles are in constant movement on the road. Therefore, for brevity of exposition, the transmission boundary of the RSU is adopted when computing the broadcast transmission rate. The accuracy of such approximation is acceptable when the headway between vehicles is small.

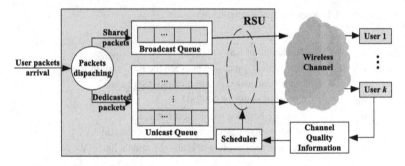

Fig. 4.2 Illustration of the traffic model

4.2.2 Traffic Model

The scheduler in the RSU contains two kinds of finite buffers, i.e., N unicast buffers and one common broadcast buffer, as shown in Fig. 4.2. In one time slot, after a content similarity analysis, the packets for broadcast transmission are pulled into the broadcast buffer, while the unicast transmission packets are stored into the corresponding unicast buffer through packets dispatching. Wireless resources are distributed by the scheduler. After a packet is sent over, it is cleared from the user queue. The traffic arrival process of the nth vehicle is modeled as a Poisson process with a mean arrival rate of λ_n. So the number of packets arriving at the RSU from the nth user during the vth time slot can be denoted by an integer $k_n(v)$, and its probability distribution is given by

$$P\{k_n(v) = m\} = \frac{(\lambda_n T_S)^m}{m!} e^{-\lambda_n T_S}, \tag{4.3}$$

where T_S denotes the length of the time slot. The size of the mth packet is exponentially distributed as follows:

$$P\{F^{(m)} \le a\} = 1 - e^{-a/\bar{F}}, \ \forall a > 0, \tag{4.4}$$

where $F^{(m)}$ is the length of the mth packet, and $\bar{F} = \mathrm{E}[F^{(m)}]$ is the average size of user packets. Without loss of generality, all the users share the same arrival rate and broadcast proportion, so that λ_n can be replaced with λ. β is defined as the broadcast proportion of each user in the total traffic. According to the stochastic decomposition theorem of the Poisson process [18], the shared and dedicated traffics also follow two Poisson processes with arrival rates $\lambda_b = \beta\lambda$ and $\lambda_u = (1 - \beta)\lambda$, respectively. Compared to the case without traffic classification, the equivalent arrival rate of the total system is reduced to $(N - 1)\beta\lambda$ approximately.

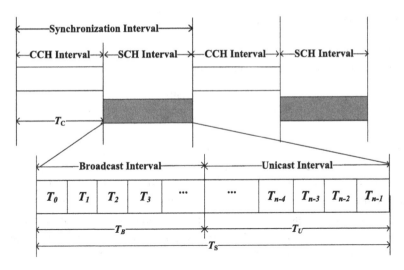

Fig. 4.3 Illustration of the TDMA frame structure

4.2.3 Analysis on Contented-Based Scheduling Scheme

4.2.3.1 Scheduling Scheme

As shown in Fig. 4.3, the channel access duration can be divided into synchronization intervals with a constant duration. One synchronization interval consists of a CCHI (T_c) and a SCHI (T_s) of the same length [19]. To work effectively, the SCH is separated into time slots of equal size. Some time slots are used for broadcast traffic, which is called the broadcast interval with a duration of T_b. Other time slots constitute unicast interval with the duration of T_u, which are dedicated to unicast traffic. The ratio of the broadcast and unicast intervals can be dynamically adjusted in accordance with the varying traffic broadcast to a variety of users. Since our interest is to improve the throughput performance of non-safety traffic, the analysis is focused only on the SCHI.

First come first serve (FCFS) is a simple allocation policy, which serves users according to the order of resource requests. All the channel resources are allocated to the user, whose data are stored at the beginning of the buffer. The users arriving later have to wait until the preceding users are served. For the sake of analysis, the Round Robin (RR) scheduling is assumed for the fair sharing of resources among users, where the channel resources are allocated to each user circularly. When the length of a time slot becomes infinitely short, all the users will be served by the scheduler simultaneously. Prior to the queueing performance analysis, the average service rate is first derived.

4.2.3.2 Average Service Rate

On the downlink of a V2I communications system, the instantaneous SNR of the nth vehicle on the kth subcarrier can be shown as

$$\Gamma_k^{(n)} = \Phi_n(l)\gamma_k^{(n)} = \frac{P_T l^{-e}||H_k^{(n)}||^2}{\sigma_N^2}, \ 1 \le k \le K, \tag{4.5}$$

where K is the number of subcarriers the nth vehicle occupies, and σ_N^2 is the noise power of the Additive white Gaussian noise (AWGN). $\Phi^{(n)}(l) = P_T l^{-e}/\sigma_N^2$ is the large-scale channel fading gain determined by the location of the vehicle, in which P_T is the transmission power of the RSU, and e is the path loss exponent. $\gamma_k^{(n)} = ||H_k^{(n)}||^2$, where $H_k^{(n)}$ denotes the fast fading channel gain of the kth subcarrier of the nth vehicle, modeled as $H_k^{(n)} \sim CN(0, \sigma^2)^1$ with $\sigma^2 = E[|H_k^{(n)}|^2]$. Correspondingly, the achievable data rate of the kth subcarrier of the nth vehicle is given by

$$C_k^{(n)} = W_k \log_2(1 + \Gamma_k^{(n)}), \tag{4.6}$$

where W_k is the bandwidth of each subcarrier. $C_k^{(n)}$ is location-dependent and time-varying. Then, $\bar{C}_k^{(n)} = E_L[E_T[C_k^{(n)}]]$ is used to denote the location and time-averaged transmission rate on the kth subcarrier of the nth vehicle, where $E_L[\cdot]$ and $E_T[\cdot]$ indicate the expectations of the location and time, respectively. Next, the average transmission rate can be given by

$$
\begin{aligned}
\bar{C}_k^{(n)} &= E_L[E_T[C_k^{(n)}]] \\
&= E_L[E_T[W_k \log_2(1 + \Gamma_k^{(n)})]] \\
&= W_k E_L[E_T[\log_2(1 + \Phi^{(n)}(l)\gamma_k^{(n)})]] \\
&= W_k \int_{\sqrt{d^2+h^2}}^{\sqrt{d^2+h^2+R^2}} \int_0^{+\infty} \log_2\left(1 + \frac{P_T l^{-\alpha}}{\sigma_N^2}\gamma_k^{(n)}\right) \\
&\quad \cdot f(\gamma_k^{(n)})f(l)d\gamma_k^{(n)}dl,
\end{aligned}
\tag{4.7}
$$

where $f(\gamma_k^{(n)})$ is the probability density function of fast fading $\gamma_k^{(n)}$. When the bandwidth per subcarrier is narrow enough, the channel in each subcarrier can be regarded as a flat Rayleigh fading one. Then, the fading coefficient follows an exponential distribution as follows:

$$f(\gamma_k^{(n)}) = \exp(-\gamma_k^{(n)}), \ 1 \le k \le K. \tag{4.8}$$

Next, the total average transmission rate of the RSU for the nth vehicle which occupies all the subcarriers is given by

$$\bar{C}^{(n)} = \sum_{k=1}^{K} \bar{C}_k^{(n)}. \tag{4.9}$$

Vehicles in the system may undergo the same communications process when they pass by the RSU. That is, they start transmission once entering the coverage area of the RSU, and stop transmission when leaving the coverage area. They experience the same location and similar channel state variation. So, we can use $\bar{C}$ in lieu of $\bar{C}^{(n)}$ to express the average statistical transmission rate of an arbitrary vehicle. The average unicast transmission rate of the RSU for each vehicle can be given by

$$\bar{C}_u = \frac{T_u}{T_u + T_b} \bar{C}. \tag{4.10}$$

Similarly, the average broadcast transmission rate can be expressed by

$$\bar{C}_b = \frac{T_b}{T_u + T_b} \bar{C}', \tag{4.11}$$

and

$$\bar{C}' = \sum_{k=1}^{K} W_k \int_{0}^{+\infty} \log_2(1 + \frac{P_T L^{-e}}{\sigma_N^2} \gamma_k^{(n)}) f(\gamma_k^{(n)}) d\gamma_k^{(n)}, \tag{4.12}$$

where $L = \sqrt{R_0^2 + d^2}$.

Therefore, the average service rates of unicast and broadcast can be computed, respectively, i.e.,

$$u_u = \left[\frac{F}{\bar{C}_u} \right]^{-1} = \frac{\bar{C}_u}{F}, \tag{4.13}$$

$$u_b = \left[\frac{F}{\bar{C}_b} \right]^{-1} = \frac{\bar{C}_b}{F}. \tag{4.14}$$

4.2.3.3 Queueing Performance Analysis

Considering that the traffic arrival process follows a Poisson distribution, both the broadcast and the unicast queueing systems can be modeled as an M/G/1 system.

(1) Unicast Transmission

The unicast transmission mode can be modeled as a Processor sharing (PS) queueing system, when the RR algorithm is employed to schedule vehicles [20]. The average queue length and the delay of each packet in the M/G/1 system are consistent with those of M/M/1-PS [20]. According to existing results on the PS model with a state-dependent service rate [21, 22], the average queue length, delay and throughput of the unicast system can be given by

$$\bar{Q}_u = \frac{\rho_u}{1 - \rho_u} = \frac{\lambda_u}{\mu_u - \lambda_u}, \tag{4.15}$$

$$\bar{\tau}_u = \frac{\bar{Q}_u}{\lambda_u} = \frac{1}{\mu_u - \lambda_u}, \tag{4.16}$$

$$\bar{\xi}_u = \bar{F}/\bar{\tau}_u = \bar{F}[\mu_u - \lambda_u], \tag{4.17}$$

where $\rho_u = \lambda_u/\mu_u$, which represents the unicast traffic load of the network.

(2) Broadcast Transmission

The broadcast queueing system behaves like an M/G/1-FCFS system, when the FCFS algorithm is employed to schedule vehicles. Thus, the average queue length and delay are influenced by the distribution of the service time. A parameter W is introduced which is related to the instantaneous service time T and the standard deviation σ_T of the shared traffics, with $W = \frac{1}{2}[1 + (\sigma_T/T)^2]$ and $T = F/C_u$ [20]. Then, the average queue length, delay and throughput of the broadcast system can be given by

$$\bar{Q}_b = \rho_b + \frac{\rho_b^2 W}{1 - \rho_b}, \tag{4.18}$$

$$\bar{\tau}_b = \frac{\rho_b^2 W}{1 - \rho_b}, \tag{4.19}$$

$$\bar{\xi}_b = \bar{F}/\bar{\tau}_b = \bar{F}(1 - \rho_b)/\rho_b T_s W. \tag{4.20}$$

where $\rho_b = \lambda_b/\mu_b$, which represents the broadcast traffic load of the network.

4.2.4   Performance Analysis

In this section, simulations are carried out to verify the theoretical results and to evaluate the performance of the proposed scheme. The RSU usually has high processing capabilities, so the time for analyzing and classifying traffic is negligible compared with that of data transmission. α is used to denote the broadcast proportion of the channel resources, and is fixed to 0.4. The average packet arrival rate of each vehicle is $6 \, s^{-1}$. Detailed simulation parameters are listed in Table 4.2.

Table 4.2 Simulation parameters

Parameter	Value
Carrier frequency	2 GHz
Bandwidth	10 MHz
Radio range	300 m
MAC scheme	TDMA
Channel model	Rayleigh Fading
Slot length	1 ms
Proportion of broadcast resources(α)	0.4
Arrival process	Poison process
Average arrival rate	$6 s^{-1}$
Average packet length	300 kbit
Vehicle speed	15 m/s

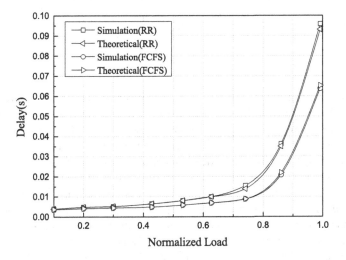

Fig. 4.4 Delay comparison of the theoretical and simulated results

As shown in Fig. 4.4, it is obvious that the analytical results agree well with their simulated counterparts. Meanwhile, FCFS scheduling scheme has better delay performance in comparison with the RR scheduling scheme.

In order to investigate the effect of the ratio between the shared and broadcast traffic on the system performance, four cases are considered, i.e.,

– *Case 1*: No traffic classification, unicast transmission for all data traffic, set as the reference scheme;
– *Case 2*: Traffic classification is employed, proportion β of the traffic is broadcast, the remaining portion is unicast, and $\alpha > \beta = 0.2$;
– *Case 3*: Traffic classification is employed, and $\alpha = \beta = 0.4$;
– *Case 4*: Traffic classification is employed, and $\alpha < \beta = 0.6$.

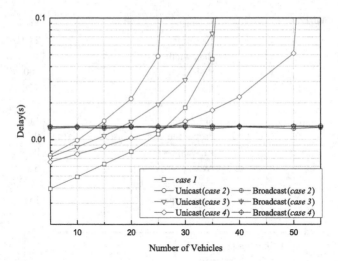

Fig. 4.5 Delay comparison

The delay performance of the four cases is given in Fig. 4.5. For shared traffic, the delay rises slightly with the increasing of β due to the growing shared traffic. However, it remains approximately the same irrespective of the number of vehicles in the system through reusing the network resources, which is the exact advantage of the broadcast transmission mode. Compared with the shared traffic, it is more complex about the cases of the dedicated traffic. In *Case 1*, all traffic is transmitted without difference using all the channel resources provided by the RR scheduler. Unlike *Case 1*, the unicast traffic can only be transmitted during the unicast slots that account for 60 % of the channel resources available in *Cases 2–4*. Thus, if one packet does not complete transmission in one frame, its entire transmission delay may be added to the next broadcast interval. This kind of delay can be mitigated when the length of the delayed frame becomes shorter and the average packet length becomes smaller, although the delay cannot be eliminated completely.

The delay performance of *Case 2* is always inferior to that of *Case 1*. Apart from the additional delay mentioned above, there is another reason that 80 % of the dedicated traffic is transmitted on 60 % of the resources, so that the equivalent unicast arrival rate per unicast scheduling slot increases by 30 % compared with the reference case. In *Case 3*, a small gap also exists compared with the reference in terms of delay when the number of the vehicles is under 35, although the equivalent unicast arrival rate per unicast scheduling slot is the same as that of the reference case. When $\alpha < \beta$ in *Case 4*, the problem of the short unicast interval also exists. However, the equivalent unicast arrival rate is smaller. Therefore, compared with *Case 1*, the delay performance improves when the number of vehicles is creased to 25 and over.

Figure 4.6 shows the average throughput performance of each user, which includes both the unicast and broadcast performances. When employing the traffic

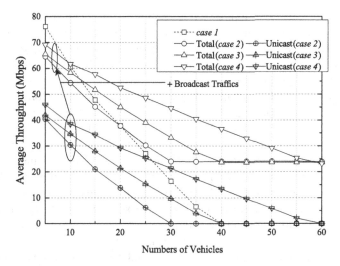

Fig. 4.6 Average throughput comparison

classification scheme, the throughputs of *Case 2, 3* and *4* are higher than that of *Case 1*, when the number of vehicles exceeds 20, 13, and 10, respectively. Moreover, the larger the β is, the greater the increase, and the descent slope of the throughput in *Case 1* is the fastest.

4.3 Bipartite Graph-Based Cooperative Resource Allocation Mechanism

VE close to the base station may enjoy a favorable channel quality resulting in high data rates. However, the others far away from the BS can only have much lower data rates due to poor radio links. To tackle this challenge, a cooperative relaying mechanism among neighboring vehicles is desired to be established for V2V communications. Due to the broadcast nature of the wireless channel, intermediate nodes are able to receive the transmitted signals, but only some of them take part in cooperative relaying. The problem we face is how to determine non-cooperative or cooperative transmission for each VE, and how to schedule VEs in proximity to the BS to help others. In this section, we investigate these cooperative relaying problems in cellular-based vehicular networks with V2V communications by proposing a new graph-based approach [23]. Most existing graph-based resource scheduling methods fall under two categories, i.e., (1) graph coloring [24–27]; and (2) maximum weighted matching (MWM) in a weighted bipartite graph [28]. This paper focuses on the latter one, which is relatively less well-investigated.

In this section, we employ graph theory to formulate the problem of scheduling the V2I and V2V links in vehicular networks. Due to the tree structure of a relay network, a feasible approach is to solve the spanning tree of a complete graph, which contains all the possible links in the network. However, this brute-force approach results in intractable computational complexity owing to exhaustive search. Therefore, we propose a BG bipartite graph (BG)-based scheduling scheme, consisting of the following three stages: (1) construct the weighted bipartite graph; (2) solve MWM; and (3) optimize the number of relayed VEs. More specifically, we first construct a bipartite graph by grouping the VEs with one subset containing the 1-hop VEs, and the other subset comprising the 2-hop VEs. The edges are weighted according to the capacity of the links between VEs. Then, we use the Kuhn-Munkres (KM) algorithm to solve the MWM problem of the constructed bipartite graph. Through stages (1) and (2), one can obtain an optimized solution of the link arrangement in the vehicular network according to certain separation of the VE set. And at stage (3), a search algorithm, such as binary search or golden section search, can be employed to find the optimal separation, through repeating stages (1) and (2) [29]. The proposed BG-based scheme leads to a much lower complexity than the exhaustive search for the optimal solution, and can be demonstrated to perform extremely close to the optimal one. In addition, it provides better fairness among VEs, and can improve the data rates of the VEs under poor channel conditions.

4.3.1 System Model and Formulation

Figure 4.7 illustrates a wireless network with N VEs on an urban road for downlink transmission under consideration in this chapter. Each VE can play a different role of communications in the network. Similar to the normal UE in LTE networks, a VE has the ability to establish a direct link with the BS. Meanwhile, VEs can help each other to forward data or not, depending on their own channel conditions and network requirements. Therefore, the communications between the VEs and BS can be established either directly (i.e., 1-hop) or via 2-hop cooperation. Relaying should occur only when it can improve the end-to-end throughput or the coverage.

4.3.1.1 1-hop Communications

In infrastructure-based LTE networks, communications via a direct link between a VE and the BS, i.e., the V2I link, are based on the LTE specification [30]. The entire radio resources are divided into resource blocks (RBs) along the time and/or frequency domain. Users share all the RBs through a scheduling algorithm. For simplicity, the RR algorithm is used to evenly distribute the RBs to each V2I link. Assume that there are K_B RBs available for the downlink transmission in the network. Thus, the achievable rate of the ith V2I link in cell B is given by

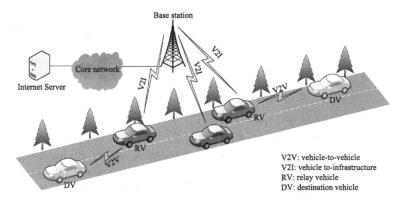

Fig. 4.7 Illustration of 1-hop and 2-hop communications with the V2I and V2V links in a vehicular network

$$\eta_{B,i} = N_I \log_2 \left(1 + \frac{\beta_{B,i} P_{B,i}}{\sum_{m=1,m \neq B}^{N_C} \beta_{m,i} P_{m,i} + \sigma_N^2} \right), \tag{4.21}$$

$$1 \leq i \leq N,$$

where N_I is the floor value of K_B/N, i.e., $N_I = \lfloor K_B/N \rfloor$, N_C is the number of cells occupying the same radio resources, $P_{m,i}$ is the transmit power from the BS of cell m on the ith V2I link, $\beta_{m,i}$ is the path loss attenuation factor from the BS of cell m to the ith V2I link user, and σ_N^2 is the noise power of the AWGN. The transmit power is assumed to be the same for all the V2I links, i.e., $P_{m,i} = P_T, 1 \leq m \leq N_C, 1 \leq i \leq N$.

4.3.1.2 2-hop Communications

When a VE is far from the BS, its data may be forwarded by another VE in proximity to the BS via out-of-band relaying communications using the IEEE 802.11p specification at a higher frequency band [31]. For ease of exposition, the VE that needs relay assistance is termed the destination vehicle (DV), while a Relay vehicle (RV) is one that can help its destination vehicle. In order to avoid a high signaling overhead and scheduling complexity, we assume that one RV can help only one DV at a time and vice versa. For 2-hop communications, data communications between the BS and the DV involve both the V2I and V2V links. Assuming that Decode-and-forward (DF) relaying is applied at the relay vehicle [32], the equivalent achievable data rate of 2-hop transmission can be given by

$$\min\{\eta_{B,i}, \eta_{i,j}\}, 1 \leq i,j \leq N, i \neq j, \tag{4.22}$$

where $\eta_{i,j}$ is the achievable data rate from the i-th relay vehicle to the j-th destination vehicle. Under the assumption that K_V radio resource units are allocated to L V2V links by RR, $\eta_{i,j}$ can be expressed by

$$\eta_{i,j} = N_S \log_2 \left(1 + \frac{\tilde{\beta}_{i,j}\tilde{P}_{i,j}}{\tilde{\sigma}_N^2}\right), 1 \le i,j \le N, i \ne j, \qquad (4.23)$$

where N_S is the floor value of K_V/L, i.e., $N_S = \lfloor K_V/L \rfloor$, $\tilde{P}_{i,j}$ is the transmit power from the i-th relay vehicle to the j-th destination vehicle, $\tilde{\beta}_{i,j}$ is the path loss attenuation factor from the i-th relay vehicle to the j-th destination vehicle, and $\tilde{\sigma}_N^2$ is the noise power of the AWGN of the V2V link. The transmit power is assumed to be the same for all the relay vehicles, i.e., $\tilde{P}_{i,j} = P_C, 1 \le i,j \le N, i \ne j$.

4.3.1.3 Problem Formulation

The links in a wireless vehicular network involving both 1-hop and 2-hop transmission lead to a complicated topology. For optimal link scheduling in vehicular communications, a graph-based problem formulation is presented in the sequel.

Based on a given network topology, we can first construct a link graph $G = (U, E)$. The vertex set U denotes the communication nodes, i.e.,

$$U = \{u_B\} \cup U' = \{u_B\} \cup \{u_i | i = 1, 2, \ldots, N\}, \qquad (4.24)$$

where u_B and u_i represent the BS and VE i, respectively, and U' is the set of the VEs. Thus, there are $|U| = N + 1$ vertexes in graph G. E is the set of all the links that may be established and that includes not only the edge set E_B of the V2I links but also the edge set E_V of V2V, i.e.,

$$E = E_B \cup E_V \qquad (4.25)$$
$$= \{e_{B,i} | i = 1, 2, \ldots, N\} \cup \{e_{i,j} | i, j = 1, 2, \ldots, N, i \ne j\},$$

where $e_{B,i}$ denotes the edge from u_B to u_i, and $e_{i,j}$ is the edge from u_i to u_j. Each edge in G is associated with a weight, dependent on the transmission capacity of the link, i.e.,

$$w(e_{B,i}) = \eta_{B,i}, e_{B,i} \in E_B, \qquad (4.26)$$
$$w(e_{i,j}) = \min\{\eta_{B,i}, \eta_{i,j}\}, e_{i,j} \in E_V. \qquad (4.27)$$

It is noted that the resources of the V2I links are first evenly allocated to all the mobile nodes either by 1-hop or by 2-hop communications. If one node helps another one as a relay vehicle, it may use the V2I resources allocated to both itself and its assisted node. The destination vehicle communicates with its relay vehicle

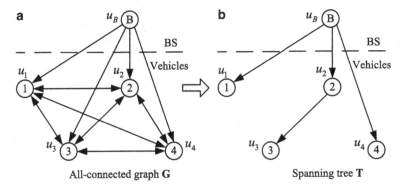

Fig. 4.8 An exemplary link graph and its spanning tree in a 4-VE network: (**a**) All-connected graph G, (**b**) Spanning tree T

using the V2V link. Hence, no data rate loss of a relay vehicle is caused due to relaying because it has been assigned extra radio resources for relaying. Moreover, if the date rate of a destination vehicle with 2-hop transmission as defined in (4.27) is higher than that with 1-hop transmission, the throughput gain can be achieved by cooperative relaying.

Graph **G** is complete since each u_i is connected with u_B by $e_{B,i}$, and any pair of nodes are linked by $e_{i,j}$. A practical and feasible 2-hop vehicular network topology is represented by a tree structure $\mathbf{T} = (\mathbf{U}, \mathbf{M})$, namely a spanning tree, which is a subgraph of **G**. The root node of **T** is u_B, i.e., the BS. There may be various spanning trees of **G**, each of which contains all the nodes in **G** but without any cycles.

For the purpose of illustration, Fig. 4.8 gives a simple example of the spanning tree of a complete graph corresponding to a network with only four VEs. We add some additional constraints for the spanning trees in vehicular networks in consideration of the power consumption and timing sequence complexity. The first constraint imposes that any VE must connect to its transmitter, being either the BS or a relay vehicle. One destination vehicle can receive data forwarded by only one relay vehicle, which has a V2I link to the BS. The second one implies that a VE can help at most one another vehicle, acting as a relay. The last one limits the number of relaying hops, i.e., at most two hops. Corresponding to Fig. 4.8a, one possible spanning tree is shown in Fig. 4.8b. Vertexes u_1, u_2, and u_4 represent the VEs which establish direct communications with the BS using the V2I links. The destination vehicle denoted by u_3 communicates with the BS through 2-hop links, including the V2I and V2V links, with the aid of RV u_2.

Therefore, the additional constraints of a allowable spanning tree **T** of the complete graph **G** are summarized as follows:

- The in-degree of each u_i $(i = 1, 2, \ldots, N)$ is limited to be 1;
- The out-degree of each u_i $(i = 1, 2, \ldots, N)$ is no more than 1;
- The depth of the spanning tree is no more than 3.

Let $\mathbf{M}$ be the set of edges in tree $\mathbf{T}$, where $\mathbf{M} \subset \mathbf{E}$. The edges in $\mathbf{M}$ stem from both the V2I and V2V links, i.e., $\mathbf{M} = \mathbf{M}_B \cup \mathbf{M}_V$, where $\mathbf{M}_B$ and $\mathbf{M}_V$ are the subsets of $\mathbf{E}_B$ and $\mathbf{E}_V$, respectively. It is clear that the vertexes in $\mathbf{T}$ are identical to those in $\mathbf{G}$. Then, we can rewrite the vertex set as $\mathbf{U} = \{u_B\} \cup \mathbf{U}_B \cup \mathbf{U}_V$, where $\mathbf{U}_B$ is the set containing the nodes connected to u_B, and $\mathbf{U}_V$ is the set containing the nodes linked to those in $\mathbf{U}_B$. Due to the characteristics of the tree structure, the number of the edges in $\mathbf{T}$ can be easily obtained as $|\mathbf{M}| = N$. On the other hand, the number of the destination vehicles using the V2V links is $|\mathbf{M}_V| = L$.

With the given spanning tree $\mathbf{T}$, the total achievable downlink data rate of the network for all the VEs can be computed with the knowledge of the weight of each edge as follows:

$$W = W_B + W_V = \sum_{e_{B,i} \in \mathbf{M}_B} w\left(e_{B,i}\right) + \sum_{e_{i,j} \in \mathbf{M}_V} w\left(e_{i,j}\right). \tag{4.28}$$

Apparently, different topologies of $\mathbf{T}$ result in different sum rates. In other words, by properly selecting the edge set $\mathbf{M}$ of $\mathbf{T}$, the maximum sum rate (MSR) can be obtained for the network. Then, the problem can be formulated into the following optimization function:

$$\tilde{\mathbf{M}} = \arg\max_{\mathbf{M} \subset \mathbf{E}} \{W\} \tag{4.29}$$

$$= \arg\max_{\mathbf{M} \subset \mathbf{E}} \left\{ \sum_{e_{B,i} \in \mathbf{M}_B} w\left(e_{B,i}\right) + \sum_{e_{i,j} \in \mathbf{M}_V} w\left(e_{i,j}\right) \right\}.$$

To the best of our knowledge, the optimal solution to problem (9) can only be found via exhaustive search. In the network with N VEs, there are N edges in $\mathbf{E}_B$, and $N(N-1)/2$ edges in $\mathbf{E}_V$. We can calculate the number of enumerations as follows:

- First, we choose n destination VEs from N vehicles that receive data with 2-hop communications. The number of possible combinations is C_N^n. The number of destination vehicles is smaller than that of relay vehicles because one relay vehicle can forward to no more than one destination vehicle, i.e., $0 < n \leq \lfloor N/2 \rfloor$;
- Then, we select n relay vehicles from the rest $(N-n)$ VEs to help the n destination vehicles. The number of permutation is A_{N-n}^n;
- For each $0 < n \leq \lfloor N/2 \rfloor$, the number of enumerations is $C_N^n A_{N-n}^n$.

Hence, to compare all the possibilities, the number of enumerations is

$$\sum_{n=1}^{\lfloor N/2 \rfloor} C_N^n A_{N-n}^n = \sum_{n=1}^{\lfloor N/2 \rfloor} \frac{N!}{n!\,(N-2n)!}, \tag{4.30}$$

which is overwhelmingly large resulting in a nondeterministic polynomial (NP) hard problem. When there are a large number of vehicles in the network, a massive number of alternative radio links for V2V communications make such a search unacceptable due to intractable computational complexity. Therefore, we propose a low-complexity scheduling scheme to arrange the 2-hop links in the vehicular network in the following section.

4.3.2 Solution of the Optimization Problem

4.3.2.1 Construction of the Weighted Bipartite Graph

Since the VEs close to the BS have good-quality V2I links, they usually establish direct communications with the BS and are able to help forward data for other VEs. Meanwhile, in order to improve the achievable data rate, those VEs far from the BS under poor channel conditions may communicate with the BS via 2-hop communications. Denote by N_B and N_V the number of the VEs with 1-hop and 2-hop communication, respectively. We choose N_V VEs with the worst channel conditions as the destination vehicles, and the rest $N_B = N - N_V$ as 1-hop VEs, of which N_V are used as the relay vehicles to aid the selected destination vehicles. Since the 1-hop VEs may or may not help another VE, we have $N_B \geq N_V$, equally, $0 \leq N_V \leq \lfloor N/2 \rfloor$. Now we need to determine which VEs are suitable for acting as the relay vehicles to help which destination vehicles.

A weighted bipartite graph $\mathbf{G}' = (\mathbf{U}', \mathbf{E}')$ is constructed on the basis of $\mathbf{G} = (\mathbf{U}, \mathbf{E})$, where the vertexes are divided into two disjoint subsets. One subset of $\mathbf{U}_V$ is the set of the VEs selected to be 2-hop destination vehicles, whereas the other subset $\mathbf{U}_B$ is the set of VEs with 1-hop communications. It is evident that $\mathbf{U}' = \mathbf{U}_B \bigcup \mathbf{U}_V$ and $\mathbf{U}_B \bigcap \mathbf{U}_V = \varnothing$. Thus, the V2I link set is $\mathbf{M}_B = \{e_{B,i} | u_i \in \mathbf{U}_B\}$, where $\mathbf{M}_B \subset \mathbf{E}_B$. Then, we can readily obtain the following relationship:

$$|\mathbf{M}_B| = |\mathbf{U}_B| = N_B = N - N_V. \tag{4.31}$$

The edges with one endpoint in $\mathbf{U}_B$ and the other one in $\mathbf{U}_V$ comprise the set of possible V2V links, denoted by

$$\mathbf{E}'_V = \{e_{i,j} | u_i \in \mathbf{U}_B, u_i \in \mathbf{U}_V\}, \tag{4.32}$$

which is a subset of $\mathbf{E}_V$, i.e., $\mathbf{E}'_V \subset \mathbf{E}_V$. Figure 4.9a gives an example of $\mathbf{G}'$ in a network with 8 VEs. VEs u_2, u_3, and u_7 are selected as the destination vehicles, whereas the others are 1-hop VEs. Thus, there are $N_V = 3$ destination vehicles, and $L = 3$ V2V links should be established. The next stage is to solve for the optimal V2V links, namely, the maximum weighted matching of the bipartite graph $\mathbf{G}'$.

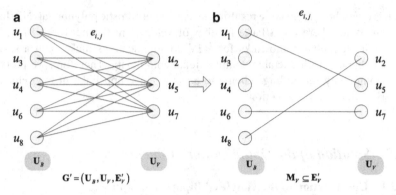

Fig. 4.9 An exemplary bipartite graph and its matching in an 8-VE network: (**a**) Weighted bipartite graph G', (**b**) A match of **G'**

4.3.2.2   Solution of Maximum Weighted Matching

A match of **G'** is denoted by $\mathbf{M}_V$ and defined as follows:

- $\mathbf{M}_V \subseteq \mathbf{E}'_V$;
- If $e_{i,j} \in \mathbf{M}_V, \forall e_{i,x \neq j} \notin \mathbf{M}_V \wedge \forall e_{y \neq i,j} \notin \mathbf{M}_V$.

Hence, $\mathbf{M}_V$ is a subset of the edges in **G'**, and no two edges in $\mathbf{M}_V$ share identical end points, as illustrated in Fig. 4.9b. Each vertex in $\mathbf{U}_B$ has no more than one connected node in $\mathbf{U}_V$. Every edge in $\mathbf{E}'_V$ is associated with weight $w(e_{i,j})$. With the following optimization objective function:

$$W_V = \sum_{e_{i,j} \in \mathbf{M}_V} w(e_{i,j}), \tag{4.33}$$

MWM satisfies that

$$\tilde{W}_V = \max \sum_{e_{i,j} \in \mathbf{M}_V} w(e_{i,j}), \tag{4.34}$$

$$\tilde{\mathbf{M}}_V = \arg \max_{\mathbf{M}_V} \{W_V\} = \arg_{\mathbf{M}_V} \{W_V = \tilde{W}_V\}. \tag{4.35}$$

We then employ the Kuhn-Munkres (KM) algorithm to solve the MWM problem of the given bipartite graph **G'** [28]. For $\mathbf{G'} = (\mathbf{U}_B, \mathbf{U}_V, \mathbf{E}'_V)$, if the cardinalities of $\mathbf{U}_B$ and $\mathbf{U}_V$ are identical, i.e., $N_B = N_V$, the bipartite graph is symmetric or asymmetric otherwise. The KM algorithm can be directly applied to a symmetric graph [28]. However, the number of 1-hop users is usually larger than that of 2-hop ones. Thus, we can expand an asymmetric bipartite graph to a symmetric one with additional $N_B - N_V$ nodes. Additional vertexes are added to set $\mathbf{U}_V$ which contains a smaller number of nodes. The added vertex set is denoted by $\mathbf{U}_V^+$, in which the

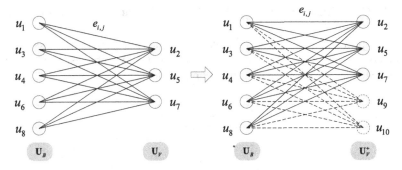

Fig. 4.10 An example of expanding an asymmetric bipartite graph of an 8-VE network

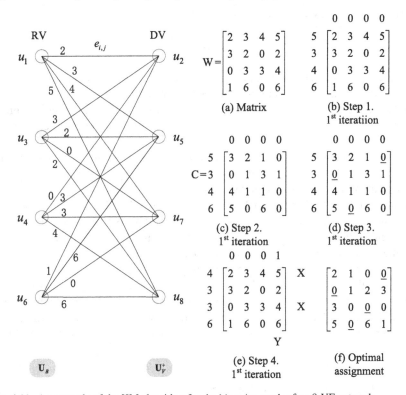

Fig. 4.11 An example of the KM algorithm for the bipartite graph of an 8-VE network

nodes are connected to those in $\mathbf{U}_B$ with zero weight. Figure 4.10 gives an example of the extended graph, where u_9 and u_{10} are two added nodes.

It is proven that the KM algorithm can always achieve MWM for bipartite graphs [28]. Next, the optimal solution to the radio link scheduling problem can be obtained by the KM algorithm. An example is also illustrated in Fig. 4.11.

4.3.2.3 Optimization on the Number of 2-hop Vehicular Equipment

From the previous two subsections, one can obtain the V2I radio links and the optimal V2V links through separating the V2I and V2V vehicles. Consequently, we can calculate the data rate of each group as follows:

$$\tilde{W}_B = \sum_{e_{B,i} \in U_B} w(e_{B,i}), \tag{4.36}$$

$$\tilde{W}_V \doteq \sum_{e_{i,j} \in \tilde{M}_V} w(e_{i,j}). \tag{4.37}$$

Then, the throughput of the entire network is given by

$$\tilde{W} = \tilde{W}_B + \tilde{W}_V. \tag{4.38}$$

In (4.38), $\tilde{W}$ is determined by parameter N_V in a certain optimization problem. In other words, given different numbers of destination vehicles (i.e., N_v), the optimized results obtained from stages 1 and 2 may not be identical. Thus, in order to achieve the optimal data rate in the network, an appropriate value of N_V needs to be solved for.

When N_V is small, the benefit of cooperative relaying is not very obvious due to the inefficient use of the out-of-band radio resources in V2V links. On the other hand, if there are too many vehicles that need help(i.e., large N_V), the amount of resources in the V2V links to each vehicle has to be reduced. Thus, the achievable rate of a vehicle with 2-hop communications is limited by its V2V link. Therefore, in the given application scenario, there exists an optimal value of N_V for the proposed scheme, which represents a good trade-off between these two effects.

The BG-based scheme proposed above requires much less computational costs. To obtain the optimal results for a vehicular downlink network, the BG-based scheme can achieve $O(N^3 \log N)$ running time, which is of polynomial complexity. The complexity of the KM algorithm is $O(N^3)$, while that of the optimization procedure of N_V is $O(\log N)$ with the binary search or golden section search [29].

4.3.3 Performance Analysis

In this section, simulation results are presented to evaluate the performance of the proposed BG-based scheduling scheme.

4.3.3.1 Simulation configuration

The main parameters and configurations of the network in our simulations are listed in Table 4.3. In our simulations, vehicles travel around an intersection, which has 500-m highways along each direction. As illustrated in Fig. 4.12, the intersection is located in the center of the cell, and a traffic light model is presented in the intersection to regulate traffic flows. The highway has five 3.5-m wide lanes, three of which enter into the intersection and two leave out of it. According to the microscopic traffic model in [33], two dynamical processes of car-following and lane-changing are considered. The car-following theory is based on the assumption that the motion of a vehicle is governed exclusively by the motion of its preceding vehicle, and is continuous in space, discrete in time, and accident-free. The random lane-changing model is adopted in our simulations, in which vehicles can change to adjacent co-directional lanes randomly.

The V2I communications use the LTE system, which transmits data via a 40 MHz bandwidth at the 2 GHz frequency with 52 dBm transmit power [30]. The V2V communications use the Wireless local area network (WLAN) 802.11p protocol, which supports the use of WLAN in the vehicle environment [31]. We adopt a 5 MHz bandwidth at the 5.9 GHz frequency with 20 dBm transmit power. Table 4.4 gives the path loss models of the V2V and V2I links used in the simulation. The path loss model of the V2I link is explained in detail in [30]. In the case of V2V links, $\tilde{\beta}_{i,j}(d_{i,j})$ is the pass loss attenuation factor at distance $d_{i,j}$, $\tilde{\beta}_{i,j}(d_0)$ is the pass loss attenuation factor at reference distance d_0, γ is the path loss exponent, and σ is

Table 4.3 Simulation parameters

Parameter	Value
Cell radius	500 m
VE number	10–100
Vehicle model	Microscopic model in [33]
Max drive peed	126 km/h (35 m/s)
Acceleration	2.6 m/s^2
Deceleration	−4.5 m/s^2
Link scheduling interval	1 s
TTI	1 ms
Thermal noise density	−174 dB/Hz
LTE configuration (V2I link)	
Carrier frequency	2 GHz
Bandwidth	40 MHz
Transmit power of BS	52 dBm for 40 MHz
DSRC configuration (V2V link)	
Carrier frequency	5.9 GHz
Bandwidth	5 MHz
VE transmit power	20 dBm for 5 MHz

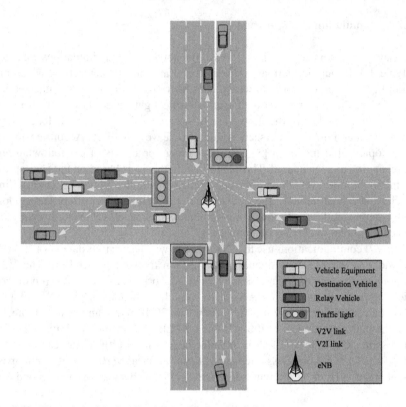

Fig. 4.12 Illustration of the intersection used in our simulations

Table 4.4 Path loss model

Link type	Path loss model
V2I link [30]	$\beta_{B,i}\left(d_{B,i}\right) = l + 37.6\log_{10}\left(\frac{d_{B,i}}{1000}\right)$, $l = 128.1 - 2\,\text{GHz}$
V2V link [34]	$\tilde{\beta}_{i,j}\left(d_{i,j}\right) = \tilde{\beta}_{i,j}\left(d_0\right) + 10\gamma\log_{10}\left(\frac{d_{i,j}}{d_0}\right) + X_\sigma$

the standard deviation of the zero-mean Gaussian variable $X_\sigma \sim \mathscr{CN}\left(0, \sigma\right)$. In the simulation, we have $d_0 = 1, \tilde{\beta}_{i,j}\left(d_0\right) = 43.9, \gamma = 2.75$, and $\sigma = 5.5$ [34].

4.3.3.2   Results and Discussions

The MSR performance of the network with exhaustive search is also evaluated for comparative purposes, i.e., the optimal solution to problem (4.29). Since the optimization of the MSR is known to be an NP problem, the number of vehicles (N) is set to no more than 40, so as to make exhaustive search feasible with the aid of pre-processing. As it is well known, when two vehicles are far from each other, cooperative relaying between them is highly unlikely. Therefore, prior to the exhaustive search, the V2V links of the vehicles, whose distance is more than 500 m apart, are excluded from the link set. The performances of the MSR algorithm with or without such pre-processing and $N = 20$ are compared, which show little difference. Then, for the cases of $N = 30$ and $N = 40$, we directly apply this pre-processing in order to obtain the performance of the MSR algorithm.

Figure 4.13 presents the cumulative distribution functions (CDFs) of the data rates with the MSR and BG-based schemes and various numbers of vehicles. It is shown that the data rates of the BG-based scheme are close to those of the optimal solution. In addition, the proposed scheme effectively decreases the distribution at the low-rate area, reducing the number of VEs suffering from low-speed transmission. For example, as can be seen from Fig. 4.13, the CDF curves of MSR are higher than that of the BG-based scheme at around 500 kbps, i.e., a very low date rate. This means that there are more users suffering from low-speed transmission.

There usually exists some degree of inaccuracy in link estimation, which may affect the performance of the V2V network. The SNR inaccuracy is modeled as a random variable with a Gaussian distribution, which has zero mean and a variance of Δ for each link. Thus, the inaccurate SNR is used for calculating the link weights. Figure 4.14 shows the date rate CDF of the BG-based scheme either with the ideal SNR or inaccurate SNR. The variance Δ is set to 0.5 or 2. It can be seen that the proposed BG-based scheme has good tolerance to the inaccuracy in SNR estimation. That is, the results with inaccurate SNRs are very close to those with ideal channel

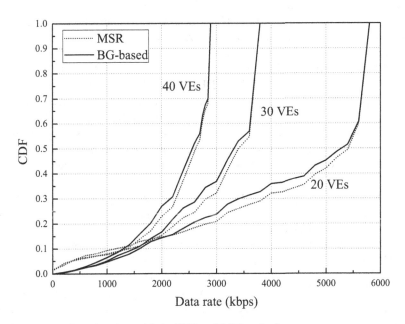

Fig. 4.13 CDFs of the data rates with the MSR and BG-based schemes

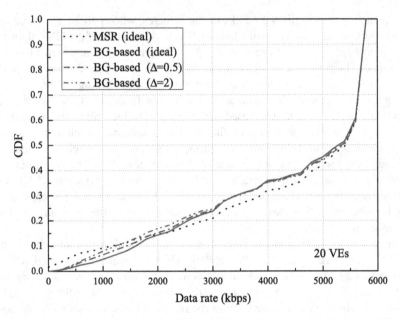

Fig. 4.14 CDFs of the data rates with the MSR and BG-based schemes with inaccurate SNRs

estimation. For instance, the average date rates are 4412 and 4387 kbps in the cases of $\Delta = 0.5$ and $\Delta = 2$, while it is 4429 kbps under perfect channel estimation without channel estimation errors. Nevertheless, with relatively serious channel estimation errors, it can be found that there are more low-rate VEs, since they cannot be correctly selected to be relayed.

To better demonstrate the performance difference, Fig. 4.15 plots the average throughout and the data rates at 5 % CDF with the MSR and BG-based schemes. Compared with the MSR scheme, there is only a slight performance loss in the average data rate, but a considerable improvement on the data rate at 5 % CDF, when using the proposed BG-based scheme. This is due to the helping-worst mechanism in our proposed scheme, which always chooses the VEs under poor channel condition as the destination vehicles with 2-hop communications. The comparison shown in Fig. 4.15 implies that the proposed scheme performs better in fairness than the MSR counterpart. Moreover, the proposed scheme has a lower computational complexity of $O(N^3 \log N)$, which is much more practical than solving the NP problem in terms of the MSR.

We also simulate the vehicular network without V2V cooperation, i.e., only using the V2I links of the LTE-advanced system. As shown in Fig. 4.16, the proposed BG-based cooperative scheme results in remarkable improvements in throughput compared to other schemes. The non-cooperative system achieves a lower data rate than the BG-based cooperative one, by using only 40 MHz bandwidth without the additional 5 MHz out-of-band frequency of DSRC. When there are more VEs in the

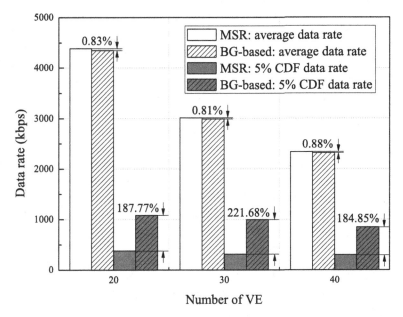

Fig. 4.15 Average data rates and the data rates at 5 % CDF with the MSR and BG-based schemes

network, the average data rate decreases due to the reduction in the amount of radio resources obtained by each VE.

To compare the non-cooperative and BG-based cooperative systems in terms of fairness, we compute their spectral efficiency (SE) as shown in Fig. 4.17. The SE is relatively higher when the number of VEs is around 40. When the vehicles on the road are sparser, their locations change rapidly with faster motion, resulting in bad channel conditions and lower SE. Whenever there are too many vehicles passing through the intersection, it will cause congestion to both the road and the wireless network, resulting in reduced data rates. In addition, under the assumption of light traffic, the radio resources of DSRC, i.e., the V2V links, are not fully utilized, since there are not always appropriate relay vehicles available to help the destination vehicles, resulting in a waste of the out-of-band resources. However, cooperative communications can achieve higher SE when there are more VEs in the network.

Finally, Fig. 4.18 plots the average number of destination vehicles N_V, while there are various numbers of VEs. This value represents the number of allocated V2V links under different levels of traffic on the road. With an increasing number of VEs, there are more destination vehicles that are selected to adopt 2-hop communications to enhance the overall throughput, attributed to the better channel quality of the V2V links due to shorter distances and potentially more appropriate VEs that can be used as the relay vehicles.

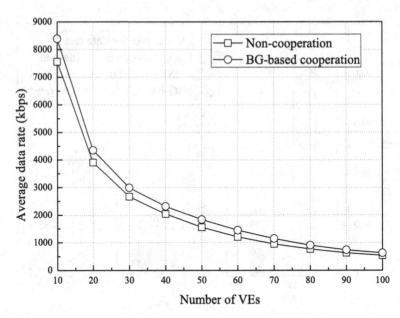

Fig. 4.16 Average data rate obtained by the non-cooperative and BG-based cooperative system with various VE numbers

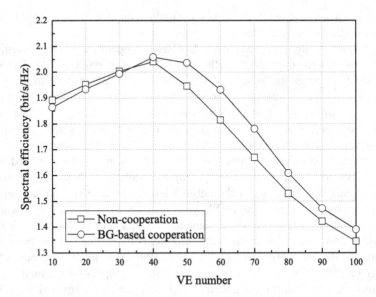

Fig. 4.17 Spectral efficiency obtained by the non-cooperative system and BG-based cooperative systems with various VE numbers

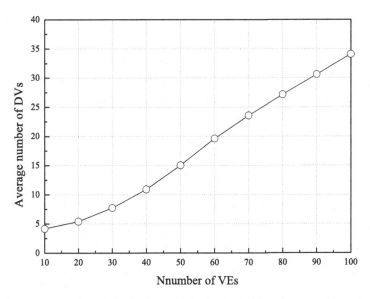

Fig. 4.18 Average number of destination vehicles in the BG-based cooperative system with various VE numbers

4.4 Summary

Following a brief review on radio resource allocation in the HetVNET, we discussed two scheduling schemes for better understanding. In the content-based scheduling scheme, the shared traffic on the service channel is extracted and broadcast so as to reduce the downlink traffic loads. The corresponding contention-free TDMA-based resource assignment scheme according to traffic classification is designed. In the BG-based cooperative scheduling scheme, the problem is formulated as graph-based optimization. Due to channel variation and mobility of VEs, the 2-hop network topology is time-varying, which can be modeled as a spanning tree based on the V2I and V2V links using graph theory. Both schemes were validated through simulation results to show their effectiveness.

References

[1] A. Damnjanovic, J. Montojo, Y. Wei, T. Ji, T. Luo, M. Vajapeyam, T. Yoo, O. Song, and D. Malladi, "A survey on 3GPP heterogeneous networks," *IEEE Wireless Commun.*, vol. 18, no. 3, pp. 10–21, Jun. 2011.
[2] J. Ghimire and C. Rosenberg, "Resource allocation, transmission coordination and user association in heterogeneous networks: A flow-based unified approach," *IEEE Trans. Wireless Commun.*, vol. 12, no. 3, pp. 1340–1351, Mar. 2013.

[3] M. Gerasimenko, D. Moltchanov, R. Florea, S. Andreev, Y. Koucheryavy, N. Himayat, S.-P. Yeh, and S. Talwar, "Cooperative radio resource management in heterogeneous cloud radio access networks," *IEEE Access*, vol. 3, pp. 397–406, 2015.

[4] S.-T. Cheng, G.-J. Horng, and C.-L. Chou, "Using cellular automata to form car society in vehicular ad hoc networks," *IEEE Trans. Intell. Transp. Syst.*, vol. 12, no. 4, pp. 1374–1384, Jun. 2011.

[5] C. E. Palazzi, M. Roccetti, and S. Ferretti, "An intervehicular communication architecture for safety and entertainment," *IEEE Trans. Intell. Transp. Syst.*, vol. 11, no. 1, pp. 90–99, Mar. 2010.

[6] H. Ilhan, I. Altunbas, and M. Uysal, "Cooperative diversity for relay-assisted inter-vehicular communication," in *Proc. IEEE Vehicular Technology Conference (VTC)*, Singapore, May. 2008, pp. 605–609.

[7] I. Krikidis, J. Thompson, and N. Goertz, "A cross-layer approach for cooperative networks," *IEEE Trans. Veh. Technol.*, vol. 57, no. 5, pp. 3257–3263, Sep. 2008.

[8] Z. Ding and K. Leung, "Cross-layer routing using cooperative transmission in vehicular ad-hoc networks," *IEEE J. Sel. Areas Commun.*, vol. 29, no. 3, pp. 571–581, Mar. 2011.

[9] W. Saad, Z. Han, A. Hjorungnes, D. Niyato, and E. Hossain, "Coalition formation games for distributed cooperation among roadside units in vehicular networks," *IEEE J. Sel. Areas Commun.*, vol. 29, no. 1, pp. 48–60, January 2011.

[10] M. Pan, P. Li, and Y. Fang, "Cooperative communication aware link scheduling for cognitive vehicular networks," *IEEE J. Sel. Areas Commun.*, vol. 30, no. 4, pp. 760–768, May 2012.

[11] J. Zhang, Q. Zhang, and W. Jia, "VC-MAC: A cooperative mac protocol in vehicular networks," *IEEE Trans. Veh. Technol.*, vol. 58, no. 3, pp. 1561–1571, Mar. 2009.

[12] T.-D. Nguyen, O. Berder, and O. Sentieys, "Energy-efficient cooperative techniques for infrastructure-to-vehicle communications," *IEEE Trans. Intell. Transp. Syst.*, vol. 12, no. 3, pp. 659–668, Sep. 2011.

[13] G. Remy, S.-M. Senouci, F. Jan, and Y. Gourhant, "LTE4V2X - impact of high mobility in highway scenarios," in *Proc. Global Information Infrastructure Symposium (GIIS)*, Da Nang, Aug. 2011, pp. 1–7.

[14] E. Yaacoub and N. Zorba, "Enhanced connectivity in vehicular ad-hoc networks via V2V communications," in *Proc. International Wireless Communications and Mobile Computing Conference (IWCMC)*, Sardinia, Jul. 2013, pp. 1654–1659.

[15] S.-Y. Pyun, D.-H. Cho, and J.-W. Son, "Downlink resource allocation scheme for smart antenna based V2V2I communication system," in *Proc. IEEE Vehicular Technology Conference (VTC)*, San Francisco, CA, 2011, pp. 1–6.

[16] Q. Wang, P. Fan, and K. Letaief, "On the joint V2I and V2V scheduling for cooperative VANETs with network coding," *IEEE Trans. Veh. Technol.*, vol. 61, no. 1, pp. 62–73, Jan. 2012.

[17] X. Xin, K. Zheng, F. Liu, H. Long, and Z. Jiang, "An efficient resource allocation scheme for vehicle-to-infrastructure communications," in *Proc. International ICST Conference on Communications and Networking in China (CHINACOM)*, Guilin, Aug 2013, pp. 40–45.

[18] L. Kleinrock, *Queueing systems, volume I: theory*. Hoboken, New Jersey: John Wiley & Sons, 1975.

[19] Q. Wang, S. Leng, H. Fu, and Y. Zhang, "An IEEE 802.11p-based multichannel MAC scheme with channel coordination for vehicular ad hoc networks," *IEEE Trans. Intell. Transp. Syst.*, vol. 13, no. 2, pp. 449–458, Jun. 2012.

[20] F. P. Kelly, *Reversibility and stochastic networks*. Cambridge, England: Cambridge University Press, 2011.

[21] S. N. M. Sakata and J. Oizumi, "Analysis of a processor shared queueing model for time sharing systems," in *Proc. 2nd Hawaii International Conference on System Sciences*, Jan. 1969, pp. 625–628.

[22] M. Sakata, S. Noguchi, and J. Oizumi, "An analysis of the M/G/1 queue under round-robin scheduling," *Operations Research*, vol. 19, no. 2, pp. 371–385, 1971.

[23] K. Zheng, F. Liu, Q. Zheng, W. Xiang, and W. Wang, "A graph-based cooperative scheduling scheme for vehicular networks," *IEEE Trans. Veh. Technol.*, vol. 62, no. 4, pp. 1450–1458, May. 2013.

[24] J. Janssen, K. Kilakos, and O. Marcotte, "Fixed preference channel assignment for cellular telephone systems," *IEEE Trans. Veh. Technol.*, vol. 48, no. 2, pp. 533–541, Mar. 1999.

[25] S. H. S. Y. Chen, N. Han and J. M. Kim, "Dynamic frequency allocation based on graph coloring and local bargaining for multi-cell WRAN system," in *Proc. Asia-Pacific Conference on Communications (APCC)*, Busan, Aug. 2006, pp. 1–5.

[26] J. Z. Yu Jung Chang, Zhifeng Tao and C. Kuo, "A graph-based approach to multi-cell OFDMA downlink resource allocation," in *IEEE Global Telecommunications Conference (GLOBECOM)*, New Orleans, LO, Nov. 2008, pp. 1–6.

[27] K. Zheng, Y. Wang, C. Lin, X. Shen, and J. Wang, "Graph-based interference coordination scheme in orthogonal frequency-division multiplexing access femtocell networks," *IET Commun.*, vol. 5, no. 17, pp. 2533–2541, Nov. 2011.

[28] H. W. Kuhn, "The Hungarian method for the assignment problem," *Naval Research Logistic Quarterly*, vol. 2, no. 1–2, pp. 83–97, 1955.

[29] D. Knuth, *The art of computer programming*. Boston: Addison-Wesley Professional, Mar. 2011.

[30] "Evolved universal terrestrial radio access (E-UTRA); further advancements for e-utra physical layer aspects (release 9)," Tech. Rep. 36.814 V9.0.0, Mar. 2010.

[31] "IEEE Standard for Information Technology– Local and Metropolitan Area Networks– Specific Requirements– Part 11: Wireless LAN Medium Access Control (MAC) and Physical Layer (PHY) Specifications Amendment 6: Wireless Access in Vehicular Environments," *IEEE Std 802.11p-2010 (Amendment to IEEE Std 802.11-2007 as amended by IEEE Std 802.11k-2008, IEEE Std 802.11r-2008, IEEE Std 802.11y-2008, IEEE Std 802.11n-2009, and IEEE Std 802.11w-2009)*, pp. 1–51, Jul. 2010.

[32] J. Laneman, D. Tse, and G. W. Wornell, "Cooperative diversity in wireless networks: Efficient protocols and outage behavior," *IEEE Trans. Inf. Theory*, vol. 50, no. 12, pp. 3062–3080, Dec. 2004.

[33] S. Krauß, "Microscopic modeling of traffic flow: Investigation of collision free vehicle dynamics," Ph.D. dissertation, Universitat zu Koln., 1998.

[34] L. Cheng, Henty, Benjamin, Stancil, D. D., F. Bai, and P. Mudalige, "A fully mobile, GPS enabled, vehicle-to-vehicle measurement platform for characterization of the 5.9 GHz DSRC channel," in *Proc. IEEE Antennas and Propagation Society International Symposium*, Honolulu, HI, Jun. 2007, pp. 2005–2008.

Chapter 5
Conclusion and Outlook

5.1 Conclusion

This monograph aims to elaborate that wireless communications networks can be utilized to provide ubiquitous ITS services with guaranteed QoS. The requirements of safety and non-safety services are summarized and compared. Based on the proposed framework of the HetVNET, current wireless networking technologies for vehicular communications are discussed in detail. The discussions on channel access and cooperative resource allocation mechanisms emphasize on the feasibility and efficiency of the HetVNET. It is concluded that the HetVNET is the most suited solution for supporting vehicular services, which requires collaboration between various communications infrastructure.

5.2 Future Research Directions

This section presents some future research directions for HetVNETs, especially those closely related to heterogeneity. Addressing these open issues is vital to alleviating the restrictions imposed by heterogeneity.

Inter-system handover: Since an HetVNET consists of various wireless networks, e.g., WCDMA, LTE, and DSRC, vehicular users may frequently switch among different networks thanks to their fast movement. It is desired that a vehicle maintains constant connection with the most suitable network. Handover is imperative to achieve seamless transmission in the HetVNET. Traditional handover mechanisms for cellular networks are mostly centralized, which are not well suited for the hybrid-distributed vehicular architecture. Also, the handover decision relies usually on a single threshold, which is influenced by a number of factors such as the network load, receiving signal strength, and channel conditions. However, there lacks an

© The Author(s) 2016
K. Zheng et al., *Heterogeneous Vehicular Networks*, SpringerBriefs in Electrical and Computer Engineering, DOI 10.1007/978-3-319-25622-1_5

appropriate model for mapping these parameters to a threshold value. Furthermore, the handover of vehicular users is more frequent than cellular users, resulting in an excessive signaling overhead. Therefore, the main challenge in designing an effective handover strategy for HetVNETs is to strike an elegant trade-off among the QoS requirements, implementation complexity, and signaling overhead.

Big Data: All participants in an ITS act as data generators, yielding huge volumes of data, e.g., beacon messages and warning messages. For instance, most commuters may like to socialize with their peers or watch popular movies in the car or bus during the long and boring commute, which may generate huge volumes of requested data. With millions of miles of roads, millions of vehicles as well as drivers collecting data over the years, the sheer number of data points is enormous. Thus, how to exploit this big data in HetVNETs has drawn much attention. However, the methods, models, and algorithms for conventional big data applications may not work well for HetVNETs. In general, big data are physically and logically decentralized, but virtually centralized [1]. In order to achieve an elegant balance between information processing and data transmission, advanced data processing and mining techniques are required to find, collect, aggregate, process, and analyze information in HetVNETs. There is much more work to be done.

Cooperation: Due to vehicle mobility, wireless links for vehicular communications are unreliable and of limited capacity. Thus, minimizing end-to-end latency and maximizing throughput are key issues in HetVNETs. Spatial diversity has been shown to be effective in enhancing energy efficiency and improving spectral efficiency in vehicular networks [2–4]. However, the multiple antennas technique is not employed in DSRC, and equipping vehicle nodes with multiple antennas may not always be practical. As an alternative solution, cooperative communications can reap the benefits of spatial diversity gains without having to install multiple antennas on each vehicle nodes. For example, due to the instability of the wireless channel, the data volume downloaded by an individual vehicle per drive-through is rather limited. In order to solve this problem, a cooperative drive-through Internet scheme, dubbed ChainCluster, is proposed to select appropriate vehicles to form a linear cluster on the highway [5]. The cluster members then cooperatively download and share the same content information, increasing the probability of successful content download. Current studies have shown that: (1) cognitive radio technology provides good opportunities for cooperative communications; (2) the performance of link scheduling with an appropriately selected transmission mode is better than relying purely on one single transmission mode; and (3) cooperative MIMO techniques provide attractive benefits for vehicular networks [2, 6]. Schemes such as link adaptation, relay selection, and radio resource management in cooperative communications are important for improving system performance. The optimization problem in cooperative vehicular communications is usually NP-hard and computationally intractable. The main issue is how to balance between performance and complexity.

Cross-layer design: HetVNETs are expected to support a wide range of safety and non-safety related services such as web browsing, file transfer, and video streaming. As opposed to traditional wireless and wired environments, the highly dynamic vehicular environment poses new challenges. For example, the communications channel is more open to unpredictability, and the connectivity among vehicles is easy to break. Hence, stringent and diversified QoS requirements of ITS services are difficult to be met by traditional layered designs. Correspondingly, there has been increased interest in exploiting the interaction among various layers of the protocol stack for performance enhancement [7–10]. The main challenge is how to design the upper layer functions based on the feedback from lower layers. At the same time, the implementation complexity of the system needs to be taken into account.

Vehicular Cloud Networking (VCN): With rapid evolution of computing and Communications technologies, the vehicles with powerful computing abilities are advocated to be service providers rather than just service recipient. As a result, the concept of Vehicular cloud computing (VCC) has been proposed, which makes join use of computation, communications, and storage resources in vehicle equipment [11], e.g., on-board computer/communications devices or mobile user equipment arrived by passengers. In general, services in a VCC system can be divided into four types according to their functions, i.e., Network-as-a-Service (NaaS), Storage-as-a-Service (StaaS), Sensing-as-a-Service (SaaS), and Computation-as-a-Service (CaaS). Different from the traditional cloud computing system, the VCC system has its unique features [12], e.g., the variation of available computational resources in Vehicular clouds (VCs). Due to the uncertainty of vehicle behaviors, e.g., vehicles may randomly join or leave a VC, the resources in the VC are time varying. Another obvious feature is the heterogeneity of VC resources. Vehicles manufactured by different vendors have different computational resources. Therefore, there are a large number of problems in vehicular cloud networking that remain to be solved.

References

[1] J. Fiosina, J. P. Müller, and M. Fiosins, "Big data processing and mining for next generation intelligent transportation systems," *Jurnal Teknologi*, vol. 63, no. 3, 2013.

[2] T.-D. Nguyen, O. Berder, and O. Sentieys, "Energy-efficient cooperative techniques for infrastructure-to-vehicle communications," *IEEE Trans. Intell. Transp. Syst.*, vol. 12, no. 3, pp. 659–668, Sep. 2011.

[3] Z. Ding and K. Leung, "Cross-layer routing using cooperative transmission in vehicular ad-hoc networks," *IEEE J. Sel. Areas Commun.*, vol. 29, no. 3, pp. 571–581, Mar. 2011.

[4] K. Zheng, F. Liu, Q. Zheng, W. Xiang, and W. Wang, "A graph-based cooperative scheduling scheme for vehicular networks," *IEEE Trans. Veh. Technol.*, vol. 62, no. 4, pp. 1450–1458, May. 2013.

[5] H. Zhou, B. Liu, T. Luan, F. Hou, L. Gui, Y. Li, Q. Yu, and X. Shen, "Chaincluster: Engineering a cooperative content distribution framework for highway vehicular communications," *IEEE Trans. Intell. Transp. Syst.*, vol. 15, no. 6, pp. 2644–2657, Dec. 2014.

[6] M. Pan, P. Li, and Y. Fang, "Cooperative communication aware link scheduling for cognitive vehicular networks," *IEEE J. Sel. Areas Commun.*, vol. 30, no. 4, pp. 760–768, May 2012.

[7] K. Zheng, Y. Wang, L. Lei, and W. Wang, "Cross-layer queuing analysis on multihop relaying networks with adaptive modulation and coding," *IET Commun.*, vol. 4, no. 3, pp. 295–302, Feb. 2010.

[8] J. Nzouonta, N. Rajgure, G. Wang, and C. Borcea, "VANET routing on city roads using real-time vehicular traffic information," *IEEE Trans. Veh. Technol.*, vol. 58, no. 7, pp. 3609–3626, Feb. 2009.

[9] K. Lee, S.-H. Lee, R. Cheung, U. Lee, and M. Gerla, "First experience with cartorrent in a real vehicular ad hoc network testbed," in *Proc. Mobile Networking for Vehicular Environments*, Anchorage, AK, May 2007, pp. 109–114.

[10] V. Cabrera, F. Ros, and P. Ruiz, "Simulation-based study of common issues in VANET routing protocols," in *Proc. IEEE Vehicular Technology Conference (VTC)*, Barcelona, Apr. 2009, pp. 1–5.

[11] L. Gu, D. Zeng, and S. Guo, "Vehicular cloud computing: A survey," in *Proc. IEEE Globecom Workshops (GC Wkshps)*, Atlanta, GA, Dec. 2013, pp. 403–407.

[12] K. Zheng, H. Meng, P. Chatzimisios, L. Lei, and X. Shen, "An SMDP-based resource allocation in vehicular cloud computing systems," *IEEE Trans. Ind. Electron.*, vol. PP, no. 99, pp. 1–1, 2015, DOI: 10.1109/TIE.2015.2482119.

Printed in the United States
By Bookmasters